HANDBOOK OF ENVIRONMENTAL SCIENCE

About the Book

Environmental science is a multidisciplinary academic field that integrates physical, biological and information sciences to the study of the environment, and the solution of environmental problems. Environmental science provides an integrated, quantitative, and interdisciplinary approach to the study of environmental systems. Related areas of study include environmental studies and environmental engineering. Environmental studies incorporates more of the social sciences for understanding human relationships, perceptions and policies towards the environment.

About the Author

***Dr. Bhawna Garg,** has vast experience of 20+ years in teaching in various government and private organisations, along with throughout first class academic record with Gold Medal in MSc. Statistics,having excellent computer proficiency in the related areas of the various subjects of Statistics like SPSS. Giving guidance to the M.Phil and PHd students of various universities. Taking the responsibilities of publication of Pranjana international journal of management awareness and research development cell.She has authored 9 Books in the field of Statistics, Entrepreneurship, HR and Neural Networks etc. and 30 Research Papers have been published in reputed research journals National as well as International. Several books and research papers have been published abroad.Life member of "Society for reliability engineering, quality and operations management (SREQOM)"Delhi. Member of editorial board of different national and international journals.Phd guide to guide students for Phd JJTU University Rajasthan and Asia Pacific university Udaipur.*

HANDBOOK OF ENVIRONMENTAL SCIENCE

DR. BHAWNA GARG

AMIGA PRESS INC.

DELHI (INDIA)

AMIGA PRESS INC.
H.O.: 28-A, Indra Park, Chander Nagar
Delhi-110051 (India)
Tel.: 011-22504028, +91-98736-57000
E-mail: amigapress@gmail.com
amigapub@gmail.com
Website : www.amigapress.co.in

This edition is published by Amiga Press Inc.

ISBN: 978-93-84533-45-8

First Published 2019

Printed and bound in India by Replika Press Pvt. Ltd.

Preface

Environmental science is a multidisciplinary academic field that integrates physical, biological and information sciences to the study of the environment, and the solution of environmental problems. Environmental science provides an integrated, quantitative, and interdisciplinary approach to the study of environmental systems. Related areas of study include environmental studies and environmental engineering. Environmental studies incorporates more of the social sciences for understanding human relationships, perceptions and policies towards the environment.

Environmental engineering focuses on design and technology for improving environmental quality in every aspect. Environmental scientists work on subjects like the understanding of earth processes, evaluating alternative energy systems, pollution control and mitigation, natural resource management, and the effects of global climate change. Environmental issues almost always include an interaction of physical, chemical, and biological processes.

Environmental scientists bring a systems approach to the analysis of environmental problems. Key elements of an effective environmental scientist include the ability to relate space, and time relationships as well as quantitative analysis.

Environmental science came alive as a substantive, active field of scientific investigation in the 1960s and 1970s driven by the need for a multi-disciplinary approach to analyze complex environmental problems, the arrival of substantive environmental laws requiring specific environmental protocols of investigation and the growing public awareness of a need for action in addressing environmental problems.

Events that spurred this development included the publication of Rachael Carson's landmark environmental book Silent Spring along with major environmental issues becoming very public, such as the 1969 Santa Barbara oil spill, and the Cuyahoga River of Cleveland, Ohio, "catching fire" and helped increase the visibility of environmental issues and create this new field of study. This encyclopedia will prove to be useful and informative for students, research scientists, engineers and teachers in the field of environment science and technology.

–Author

Contents

1

Introduction

ATMOSPHERE

Some of the atmosphere's profoundest mysteries are far above any possible ascent of man-carrying balloons or propellerdriven airplanes— beyond reach, even, of the twenty-five-mile sounding balloon. In the upper stratosphere there are levels warmed by the action of sun rays on ozone. Beyond the stratosphere, in the ionosphere, there are levels where solar radiation separates innumerable free ions and electrons out of the extremely rarefied air. Some traces of our earth's atmosphere, becoming ever more and more tenuous with increasing distance, probably extend out to cosmic depths of two thousand miles.

Our present knowledge of the upper stratosphere and the ionosphere beyond is mostly indirect painstakingly gleaned, that is, from such intangible phenomena as sound reflection, light reflection, radio reflection, semi-cosmic clouds, meteors, polar lights, and so on. Or, to put it plainly, our present picture of the upper atmosphere is largely limned with guesswork; and subject to constant correction as Doctor X announces a new and inexplicable observation, as Doctor Y propounds a new and scarcely intelligible theory, or as Doctor Z bobs up to confound Doctor Y. This state of affairs is likely to continue until the sounding rocket at last carries recording instruments, or possibly human observers, above the lower stratosphere into black depths beyond the blue.

Nevertheless, innumerable passive observations, as well as certain activeexperiments with sound waves and radio waves, have already revealed manyinteresting facts about the upper atmosphere.

ABOVE THE STRATOSPHERE

The colour of the sky, of course, comes from the blue light which the air molecules scatter, and at sunset the light slanting through miles and miles of dense air is predominantly of penetrating red. The duration of twilight is a key, imperfect but not entirely valueless, to the density of the upper air; for the twilight afterglow, so infused with simple and reposeful beauty, is nothing but the shining of air particles high overhead in the rays of the sunken sun. Knowing the time after sunset through which twilight persists, perhaps an hour or two, and the consequent dip of the sun below the horizon, perhaps eighteen degrees or so, scientists have calculated that air sufficiently dense for this reflection extends to a height of about fifty miles.

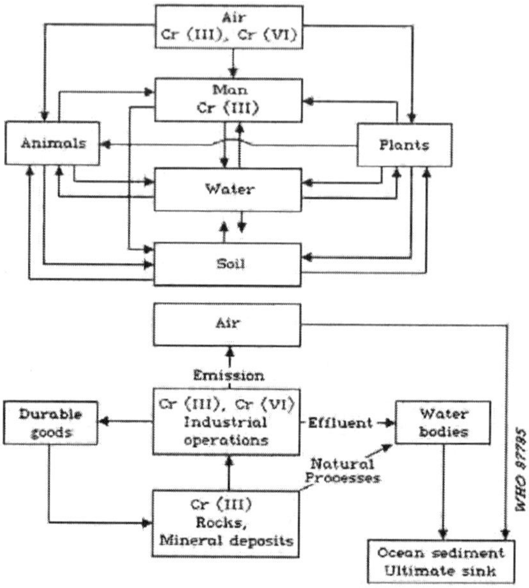

Fig. 1.1: General Structure of the Air Ocean

This fifty-mile altitude is perhaps the upper limit of air as we can perceive it and imagine it the end of the stratosphere and the beginning of the ionosphere. Also to be seen at dawn and dusk under favourable conditions is the great domed shadow of the earth itself on the upper atmosphere, rising or sinking in that quarter of the sky opposite from the sun a few minutes after sunset or before sunrise.

It is seen best in crystal-clear weather from high altitudes. The curving shadow-are the sun lights the upper atmosphere with pale rose; below it, there is only dull-gray darkness. Another sunset phenomenon that tells somewhat of the nature of the upper air is the rose-tinted glow known as the 'purple light,' seen after sunset on very clear evenings above the point where the sun disappeared. It is caused by the refraction and reflection of the longer-wave (red) light through the upper atmospheric layers around the bulge of the earth. An observer out in space (say on the moon during a lunar eclipse) would see it as a continuous ruddy aureole around the whole globe.

Though the stratosphere and the black depths above it are cloudless in the ordinary sense, clouds of a very diaphanous kind are sometimes observed floating at these upper levels. At heights of fifteen to twenty miles patches of parti-coloured filmy iridescence, called nacreous or 'mother-of-pearl' clouds, sometimes appear in the crystal-clear skies attending an invasion of cold, dry polar air after the passage of a cyclonic rain area.

These are distinct from, and much higher than, the cirriform clouds that sometimes show iridescence around their edges; in a true nacreous cloud, floating ten miles above the troposphere, the play of colour extends clear into the cloud centre. Nacreous clouds are perhaps composed of microscopic ice crystals or minute, super-cooled water droplets; and they show the existence of mid-stratospheric winds that are mostly from westerly and northwesterly in north temperate and sub-arctic latitudes, ranging in violence from flat calm to more than a hundred and fifty miles an hour. Considerably higher, marking the beginning of the ionosphere, are noctilucent clouds, occasionally seen after sunset in arctic and sub-arctic latitudes at heights around fifty to fiftyfive miles -- heights determined, of course, by triangulation from two or more ground stations situated along known base lines. Noctilucent clouds are excessively thin and wispy in appearance, and their colour shades from golden near the horizon to white and bluish-white above.

Unlike any of the lower clouds, they probably come from outer space and probably consist of excessively fine cosmic dust. Perhaps unable to penetrate the denser atmosphere below fifty miles, they always seem to hang just above this level. Appearing as faintly luminous clouds in the night sky, they are, nevertheless, actually lit by sun rays, and perhaps energized by solar ultra-violet, beyond the earth's shadow. Noctilucent clouds give evidence of superhurricane winds at the base of the ionosphere -- drifts from easterly and northerly that sometimes blow at two hundred or three hundred or even six hundred miles an hour!

In 1879, Josef Stefan discovered in old Vienna an important law of physics that might seem at first sight far removed from any usefulness to aerologists. Stefan's law is, in brief, that both the sort and intensity of light rays (both visible and invisible) emitted by any hot body are related to its temperature. The temperature of the sun is known; therefore, physicists can say how far its radiations extend into the ultra-violet. But it was found that, as we measure them on earth, solar ultra-violet rays stop short of this theoretical spectrum (which extends, with decreasing intensity, even down to X-rays). Hence something in the atmosphere must be absorbing them. That something has turned out to be ozone—the rare triatomic form of familiar oxygen.

Some slight traces of ozone occur near the earth's surface -popularly associated with pine woods and sea breezes. Measurable amounts are formed by electric sparks, and particularly by lightning flashes. But most of the atmosphere's ozone (still only a minute percentage of the other aerial gases) occurs at heights around fifteen to twenty miles—the zone of the nacreous clouds. At heights of twenty-five to thirty miles the ozone layer is more tenuous, but more important because it receives the full force of solar ultraviolet. It is a fortunate thing for all God's creatures that this great natural screen of ozone hangs up there thirty miles aloft, incredibly thin and diffused, transparent and invisible, and until recently unknown. For without it, the far ultra-violet and near X-rays shot forth from the sun would destroy all life on the sunlit earth within a few seconds.

But the high-altitude ozone screen has an important effect on the upper air itself. In the words of Haurwitz, 'ozone as a powerful absorber of solar radiation determines to a large extent the temperature of the upper atmosphere.' We have already seen that temperature may increase somewhat with elevation in the lower stratosphere. In the upper stratosphere, where ozone enters powerfully into the picture, temperatures may be much higher— say up to summer heat (around +85°F) or even near boiling (around +160°F). These apparent high temperatures have been discovered by nothing more tangible than sound waves.

SKY-REFLECTED SOUND

It is no new fact that powerful noises, such as heavy gunfire, are sometimes distinctly heard at distant places, though entirely unheard at nearer points. Back in 1666, Londoners heard the British Fleet cannonading the Dutch lineof- battle in the Channel, while the sound apparently skipped nearer places along the English coast. This circumstance puzzled Samuel Pepys and other philosophers of that day; and was not explained, in fact, until 1903, by a German named von dem Borne.

Since von dem Borne's time it has often been noticed that the sound of a loud explosion, though heard at great distances, passes completely over intervening points. This peculiar circumstance is best accounted for by supposing that there is in the upper atmosphere some sort of layer which acts in effect as a great sound reflector, and directs outgoing sound waves downward again towards the earth. In order to fit in with all the other known facts about the upper air, this reflecting 'roof' must take the form of a very warm and very diffuse upper-air layer—it is, in fact, the upper part of the ozone layer. As the velocity of sound is greater in warmer air, the outgoing sound waves travel faster in their upper portions (which encounter high temperatures first), and are hence gradually bent back to earth. The net result is the same as if they had been reflected back from the warm upper layer.

Many experiments to check the existence and height of this soundreflecting layer were performed in Germany and England soon after the First World War Whipple, for example, set up some super-sensitive microphones while British artillery was firing on an island in the lower Thames. He heard the gunfire at Birmingham, two hundred and thirty miles away; the sounds had skipped intervening places entirely and seemed to come from the upper air. Studies of the skip-distance patterns of sound from dynamite explosions, extending out to a distance of six hundred miles, point to the same general conclusion. The only way they can well be accounted for is by assuming high atmospheric temperatures at twenty to thirty miles altitude. The actual reflection summit, at about twenty-five miles, corresponds closely with what is known and assumed about the ozone layer.

The actual pattern of audibility of some great explosion, such as an ammunition dump or a shipload of TNT, is somewhat as follows: The blast is of course heard, and almost felt, out to a distance of perhaps twenty-five or fifty miles. From there to around a hundred miles is a 'skip zone,' or ring of comparative silence. From perhaps a hundred to a hundred and fifty miles extends the first zone of reflected audibility; then a second skip zone; then a second audibility zone from perhaps two hundred and fifty to three hundred miles; also, if the explosion is powerful enough, third and perhaps fourth or fifth skip and audibility zones. The sound of this battle practice was plainly audible at Santa Barbara, one hundred and twenty miles from the ships. It came in as a dull, jarring, almost soundless rumble that rattled house walls and windows like a mild earthquake.

These upper-air sound reflections are subject to curious vagaries. There are differences between eastward and westward transmission, varying with seasonal effects on the upper-wind distribution. The reflecting layer is warmer and contains more ozone in winter; it is colder and less effective in summer. The curved reflecting layer sometimes serves to intensify the downcoming sound in a peculiar way. In one case, firing with sixteen inch guns actually broke windows in a city nearly two hundred miles away!

The noisiest sound ever heard by civilized men blasted forth when a volcanic island called Krakatoa, between Sumatra and Java, blew up in 1883. The actual blasted-out material, pumice and rock, smashed and covered whole forests on islands miles away. The volcano eruption shot to a height of seventeen miles, and its black pall of concentrated dust extended out one hundred and fifty miles. An appalling ocean tidal wave rolled out, a hundred feet high, for thousands of miles, swamping hundreds of ships and drowning thirty-six thousand people.

The actual sound of this prodigious blast was faintly heard at Rodriguez Island, nearly three thousand miles away, and powerfully audible at many places more than a thousand miles distant! The atmospheric wave-disturbance, in sub-audible form, travelled several times around the earth. Also over the whole earth, carried by tropospheric and stratospheric winds during the next few months, spread thin brownish clouds of Krakatoa's dust.

LIGHTS OF THE OUTER AIR

Definite and useful indications about the upper atmosphere are given by meteors, those swift points of light, erroneously called 'shooting stars,' that streak across the starry sky every few minutes on a clear, moonless night. As these bits of cosmic débris bore into the atmosphere at speeds of ten to forty miles per second, it is no wonder that they are intensely heated by friction and soon burst into flame. If they are observed or photographed simultaneously against a fixed-star background from two or more known locations, their altitude at successive instants can be determined, and it is found that most of them begin to flame about one hundred miles or more above the ground and burn completely out before reaching the thirty-mile level. The larger and slower ones, that is, may continue down to thirty miles, but most of the smaller and faster ones burn out around fifty miles. This knowledge, of course, gives a rough idea of atmospheric density at these heights.

Meteors are seen at their best and brightest in the small hours of the morning; for at that time one is on the 'forward' side of the earth as it swings at a comfortable eighteen miles a second along its orbit, and the fiery collisions of the meteors with our outer atmosphere are more or less head-on. Perhaps ten or twenty million meteors enter the earth's atmosphere every day. When we meet a definite swarm of them, this number is vastly increased; during the Leonid swarm of 1833 they appeared as a night-long shower of celestial snow.

Most visible meteors are not much larger than pinheads (ranging, perhaps, from birdshot to buckshot in size) and are completely burned up in the higher atmosphere. But occasionally a larger meteorite, or 'fireball,' weighing several pounds or even several tons and burning brightly enough momentarily to rival the sun, crashes through to the earth's surface. Larger meteors leave trails in the upper air that gleam phosphorescently at night, and resemble lines of white smoke by day. These trails have been seen to drift rapidly before dissolving, and their motion confirms the idea that enormously high winds can occur at heights above thirty miles eteors also confirm another idea about the far upper atmosphere; for the calculated heights at which they should blaze agree with observed heights only if one assumes high temperatures at altitudes around thirty to fifty miles. This result, of course, strengthens the testimony of sound waves and ozone.

Despite their tenuous mystery, introducing new problems, the northern polar lights and their southern counterpart also offer some evidence about the upper air. In northern countries the active aurora is a sight never to be forgotten; widespreading curtains of flame appear and vanish like the draperies of an enormous stage, while arcs and bands of red and green merge into higher rays of violet and blue marching like searchlight beams across the depths of space. Carl Störmer has observed and photographed the aurora borealis many times in Norway. He finds that it occurs in two general varieties: the lower and more common types, mostly apple-green arcs and bands, flame in the earth's shadow zone at heights around sixty or seventy to one hundred miles.

The higher types, mostly blue, violet and gray-violet rays, flicker at around two hundred miles, or even occasionally as high as four hundred miles, perhaps far up where the sun is shining across the pole above the blackness of arctic night. These heights correspond roughly, with the observed heights of the ionized layers that influence radio transmission.

The spectrum of the aurora shows that the rarefied gases of these far heights are mainly oxygen and nitrogen, rather than the helium and hydrogen formerly thought to predominate in the upper air. There is, in particular, one prominent green line that results when rarefied monatomic oxygen is electrically excited. Moreover, it has been found that this diffuse green glow is always present in the night sky, over all the earth, even above the equator.

It appears to come from atomic oxygen superheated by solar ultra-violet at heights around one hundred to two hundred miles. So the natural human conception of is somewhat of an illusion. Actually about one third of the moonless night sky's luminosity comes from the stars; more than one half comes from this highaltitude oxygen glow.

The primary cause of the aurora is probably 'searchlight' beams of electrons and ions shot out of the sun—beams that rotate with the sun's rotation, somewhat after the fashion of light beams from a rotating beacon, and sweep past the earth. Sometimes, when a brilliant aurora extends visibly to temperate latitudes, this solar 'beam-excitation' can almost be seen. Thus at midnight the northern aurora may appear from the south as successive outward pulses of light, originating at the left (west) of the sky and flashing progressively towards the right (east) of the sky. These pulses were very apparent in an aurora I witnessed some years ago in New York State. It began with great hanging curtains, but changed in an hour or so to a broad, white band along the northern horizon, surmounted by radial beams of white and pale green. From observed angles and the known distance of the magnetic pole, this aurora appeared to be about a thousand miles away from me, about a thousand miles wide, and several hundred miles high. The actual speed of the light pulses in the heavens appeared to be around one hundred or two hundred miles a second; and at the earth's distance a rotating solar beam should, in fact, sweep past at about two hundred and fifty miles a second.

Probing the Ionosphere

Within recent years the aurora-haunted depths of the ionosphere have been probed daily and almost hourly by means of upward-sounding radio signals. Radio waves of course speed outward in all directions from the transmitter, along the ground and through the air, at the velocity of light. The ground part of the wave, however, usually dies out within a few score miles, 8 Encyclopaedia of Environmental Science leaving only the 'sky' wave for long-distance transmission. And the sky wave, to reach any distant place on earth, must somehow continue to bend around the bulge of our great globe. This bending of the outward-speeding sky wave is accomplished by what amounts to reflection in the ionized layers—layers where solar ultra-violet has disassociated free electrons and ions, increasing the electrical conductivity—of the earth's upper atmosphere.

Actually, just as we saw in the case of the warm ozone layer and sound, an ionized layer does not reflect radio waves; it refracts them by virtue of the fact that the outspeeding electromagnetic vibrations tend to move faster in the more intense ionization of greater heights within the 'reflecting' layer, and hence bend gradually downward again towards the ground. Nevertheless, this refraction process has the same net effect as reflection, so that one may think of the ionized layers as radioreflecting shells or 'roofs' in the upper atmosphere.

Within the last few years it has become possible to measure the height and other characteristics of the ionized layers almost at will. By means of a special transmitter, a radio 'tick,' or signal lasting less than one five-thousandth of a second, is shot upward. In something like one thousandth of a second it plunges a hundred-odd miles into space and returns to a special receiver on the ground, accurately recording the height from which it was in effect reflected—the height of an ionized layer. By this method the main reflecting layers have been probed and checked time and time again. These changes explain why a powerful broadcasting station is heard only a hundred miles or so by day, yet reaches out a thousand miles or more by night. When solar disturbances are active, some of the layer heights vary in erratic and spectacular fashion, flopping up and down by several miles in a few seconds and playing havoc with long-distance reception.

The General Picture

Before leaving the upper atmosphere it might be well to gather up some of the essential facts—the aerological elements, as a scientist would say—into one unified picture. First of all there is the matter of pressure in the air ocean above us. Pressure decreases very uniformly with increasing height. This rule never varies; pressure cannot possibly increase with height. A long and complicated formula has been deduced, called the hypsometric formula, which shows the exact variation of pressure with altitude. The pressure at a given altitude is not always the same; it is influenced somewhat by the air temperature, and also very slightly by the amount of moisture in the air.

The complete hypsometric formula must therefore include temperature and moisture terms. In general, however, pressure decreases in the lower levels, by about one inch of mercury per thousand feet of altitude (or, in the metric system, by about one hundred millibars per kilometre of altitude). At upper levels, of course, the pressure decrease with altitude is much less. At sea level the average pressure is about one thousand millibars, or about thirty inches of mercury. At about three and one half miles altitude, pressure has fallen to five hundred millibars or fifteen inches, and half of the mass of the air ocean is therefore below one. This accounts for the extraordinarily clear, transparent air characteristic of high mountain summits and high altitudes in general; it also accounts for the difficulty in breathing that most humans begin to feel at ten or fifteen thousand feet altitude. At ten miles altitude the pressure fraction has fallen to a mere tenth; at twenty miles, it is something like a hundredth; at thirty miles, something like a thousandth.

Temperature, unlike pressure, varies with height in very erratic fashion. Even in the troposphere, temperature may decrease with increasing altitude, it may remain constant, or it may even increase as one goes higher. On the average, though temperature does decrease with height in the troposphere, by the amount of about 3°F per 1000 feet or about 6°C per kilometre. But above the troposphere, throughout the lower stratosphere, temperature is fairly constant regardless of altitude. This isothermal layer extends upward to about fifteen or twenty miles in winter; and perhaps extends higher, to about twentyfive or thirty miles, in summer.

Above this isothermal layer, temperature increases until, at heights around thirty miles (maximum absorption of ultra-violet by ozone), the almost vacuous air is probably as warm as the tropics at sea level (say around + 85°F or + 30°C), or perhaps up to near-boiling heat (say around + 160°F or + 70°C). Above this warm layer there is perhaps a cooler one. Then, at still greater heights above sixty miles (maximum absorption of ultra-violet by atomic oxygen), temperature may increase to several hundred degrees—the heat of a roasting oven—at least in the daytime. But at night these far heights are probably much colder—say below zero F. And above this upper warm zone, at altitudes of two or three hundred miles, for all we know, the temperature may decrease again to something near absolute zero.

The composition of the upper air has long been a bone of contention. Most meteorological texts printed prior to 1930 or so show the atmosphere above sixty miles (one hundred kilometers) as consisting largely of hydrogen and helium. More recent theoretical views of the matter, such as those of Chapman and Milne, Gutenberg and Haurwitz, have disproved this idea completely. In the light of present knowledge it appears that the composition of the atmosphere up to at least thirty miles, and probably up to sixty miles— so far as the proportions of various gases are concerned—must be very much like that of air in the lower altitudes.

At lower levels, in any case, the air consists on the average of about 78 per cent of nitrogen, 21 per cent of oxygen, nearly 1 per cent of argon, 0.03 per cent of carbon dioxide, 0.01 per cent of hydrogen, and even more minute fractions of neon and helium. At lower levels there is also, of course, a variable amount of water vapour, ranging up to about four per cent of the air in volume—a factor of great practical importance in tropospheric weather which will be dealt with at some length. In general this water vapour decreases with height, being almost absent from the stratosphere; but even up into the ionosphere, there are probably minute traces of it.

In the upper stratosphere the proportion of helium may increase somewhat. In the ionosphere the oxygen percentage may decrease, most of what oxygen there is takes the monatomic form, and the percentage of nitrogen probably increases. Helium and hydrogen are probably not present in these black depths above the blue, warmed to roasting heat by the super-glaring sun—for if present, they would soon diffuse away into cosmic space.

COMPOSITION OF ATMOSPHERE

The most primitive man—the wildest savage—recognizes a wind when out in it, but it took a Greek philosopher to tell us what it is: air in motion. What motion is we know, at least well enough for practical purposes, but what is air? What is that invisible and odourless something we breathe and therefore call atmosphere, the thing that affords us all our weather perceptions and whose states and conditions we have learned to measure and even to foretell, and what was its origin?

ORIGIN OF THE ATMOSPHERE

To start as nearly as possible at the beginning, how and when did the earth ever get an atmosphere—its gaseous envelope? Well, there are two, and so far as we know, only two basic substances, the electron, a certain extremely minute quantity of negative electricity, plus something else, maybe; and the proton, an equally small quantity of positive electricity, plus something else, also maybe. The neutron or chargeless mass is omitted as so little is known about it.

Of the origin of these entities no one has the slightest idea. They can exist separately, in which case they are electrically very active; or variously grouped together in equal numbers, with increase of inertia but almost total loss of electric force, at least on things external.

Every such group that is stable, or even measurably durable, is a chemical element. There are no other elements, and this number is limited. Furthermore, though occurring in unequal quantities, all these elements appear to be distributed throughout the universe.

Most of those known on the earth have been found in the sun, for instance, and we believe the others are there, too, even if in such relatively small amounts as to be difficult of detection. Hence, when that other star, according to one cosmic theory, some three billion four hundred million years ago, passed so near our sun as to drag off from it by tidal action the masses that coalesced into the planets, there were present, and came off together, all the possible elements.

Fig. 1.2: Kilauea, an Air Factory

That is, the primordial material of the earth, as it was pulled off from, or out of, the sun consisted of all the elements that now make it up, so that at the very beginning of the independent existence of the earth it had, if not an actual atmosphere, at least the makings of one.

It is believed, by those who hold to this theory that as the earth mass drew together in a molten sphere the heavier and more refractory substances formed mainly the inner core, while the lighter and more volatile elements and compounds formed the rocky shell and the gaseous envelope. In this way, they claim, the earth soon had oceans and an atmosphere of some kind. However, even if all the gases, hydrogen, oxygen, nitrogen, and others, that can combine with various elements and compounds and form solids had then so com- bined, leaving neither water nor air, it is certain that before long the earth would have begun accumulating both.

It is doing so now through every volcano, every fumarole, and every bubbling spring, and must always have done so since before even the first crust began to form. But it seems unlikely that our present atmosphere actually did come entirely from the molten interior, whether by way of volcanic activity or otherwise, because volcanoes do not, so far as we know, give off free oxygen—volcanic gases captured before there has been any chance for admixture with the air show no trace of it. Furthermore this element could not exist uncombined in the presence of hydrogen and sulphur at high temperatures, both of which are abundant in volcanic vapors.

But we have oxygen; where did it come from? It is known that highly developed green plants extract, under the stimulus of light, a great deal of oxygen from carbon dioxide, a gas abundantly emitted by volcanos. But most primitive plants do not, they consume it; hence this promising source of free oxygen appears, on close examination, to be quite uncertain. However, some forms of lower life thrive in the absence of free oxygen and yet, in the presence of light, evolve it from certain of its compounds, especially water.

Again, lightning, which must have occurred from the beginning, frees a little oxygen from water; and so also may ultra-violet light. There have been, then, continuously active means of obtaining free oxygen from its compounds since the beginning of the world. Nevertheless, it seems most likely that at the beginning there was more oxygen present than was necessary to use up the free hydrogen and to combine with the available surface materials—enough to do all this and to have a goodly amount left over as free oxygen of the atmosphere.

Presumably, too, there were present other primitive gases, especially nitrogen, argon, carbon dioxide and water vapour. But whatever the primitive state of the atmosphere when the earth first formed, it may be regarded as practically certain that during the whole of the three billion four hundred million years since that time it has been continuously depleted by combination with many things in and of the crust, and also as continuously repleted by their decomposition. It is always changing, but, except in respect to water vapour, the change is so small in comparison with the whole that we are not ordinarily aware of it.

It has been argued that the could not have retained an atmosphere when molten, or even when dull red. But this is true of only the lightest two gases, hydrogen and helium, and of them only if no other gases were present in large amounts. A deep atmosphere of water vapour, for instance, would catch any light gas that might leave the earth beneath with an escaping velocity. It could escape only if the outer portions of the atmosphere also were quite hot.

At any rate, as soon as a crust, however thin, formed over the earth the supply of heat from beneath was so reduced (the crust being a good insulator) that the upper air necessarily became cool enough to retain the lightest gases. Also water must soon have begun gathering on the surface. Even if, up to this stage, all helium and free hydrogen had been driven wholly away from the earth, a condition that seems unlikely, there has been since then, and still is, abundant opportunity to accumulate both of them—hydrogen from volcanos, and helium from radioactive materials everywhere.

Presumably, therefore, the atmosphere is primitive in part—pulled off from the sun with all the other elements—and in part, at least, certainly regenerated inasmuch as every volcano is an active air factory.

Apparent Simplicity of air

Presumably every one usually thinks of air as being just air, a homogeneous and single thing. Many of us always think of it that way, as the ancients did, when we think of it at all. Indeed so far as its behaviour and most of its physical properties are concerned it shows no obvious complexity. It is just air in motion that is responsible for a thousand familiar things from the stir of a leaf to the wreck of a house; and just air that floats the balloon and sustains the aeroplane. In all these matters it usually is quite satisfactory to regard the air as the single substance it ordinarily seems to be.

Evidence of Complexity

This apparent oneness of air does not, however, extend to all physical processes. When we try to liquefy it, for instance, evidences of its complexity soon become amazingly conspicuous. If untreated air is forced through the cooling coils they quickly become choked with ice; and they still clog up when even the driest air is used, if nothing but the water vapour has been removed— this time with solid carbon dioxide. Then, too, the liquid air itself shows abundant evidences that it is a mixture and not a simple substance like water. What then are the known constituents of the atmosphere, and how and when were they discovered?

Water Vapour

The earliest considerations of the composition of the air that have come down to us are those of the Greeks. In their speculative philosophy on the composition of objects, they considered air to be one of the four elements (fire, earth and water being the other three) that, singly or variously combined, make up all substances.

From this it might seem that these Greek philosophers regarded the atmosphere as strictly a single thing—the "element" air. Yet it appears that by "air" they meant anything gaseous, and not necessarily the atmosphere. At any rate, Aristotle 250 years B.C., says very distinctly, in his work on meteorology, that cloud and rain are caused by condensation from the atmosphere of water vapour that had gotten there by the evaporation of water at the surface of the earth. He thus makes it very clear that the air consists of at least two things, and that water vapour is one of them.

Water vapour, then, was the first constituent of the atmosphere to be explicitly recognized. Aristotle mentions it, but it is not certain that this discovery was original with him. However, his is the earliest record we have of it, and for that reason, there being no evidence to the contrary, we regard him as one of the discoverers the earliest one of the constituents of the air.

DELAY OF FURTHER DISCOVERIES

For more than 22 centuries, therefore, and perhaps for much longer, it has been known that the air we breathe consists of at least two things, water vapour and whatever is left after the water is removed. And for more than two thousand years after the days of Aristotle this is all that was known about its composition. Indeed it was practically impossible to push our knowledge of the atmosphere any farther without something of the facilities and methods of the modern laboratory, nor before there had been acquired—very slowly and tediously it was—a fair concept of chemical elements and pure substances. Not until the beginning, then, of the 18th century was it reasonably possible for any constituent of the air to be discovered in addition to water vapour, nor indeed was any discovered until long after that. The chief obstacles that prevented such discovery for more than a hundred years after enough advance for that purpose had been made in laboratory technique, for that had been adequate from the beginning of the 17th century, were:

- The fixed idea that all gases are alike, all just air, and that any differences between various samples are due only to greater or less modifications of one and the same thing.
- The completely misleading and faulty concept that flame or combustion is the escape of something, phlogiston they called it, from within the burning object.
- The failure to recognize that change of weight incident to strong heating, during combustion, or under any other circumstances, was a matter of importance or had any scientific significance whatever.

A century before any constituent of the atmosphere, except of course water vapour, was recognized and collected in an approximately pure state, and while the faulty notions just listed were still prevalent, two people, working entirely independently, came near to finding one or more of its elements. The first of these was Robert Boyle a wealthy bachelor, chemist and theologian; discoverer of the fact, known as "Boyle's Law," that doubling the pressure on a gas reduces its volume by one-half, of course for the same temperature. It seems very probable that Boyle would have discovered some of the constituents of the air if he had carried to completion certain experiments that he definitely listed. But there is nothing in his voluminous writings to show that he ever got them beyond the paper stage, despite the fact that, so long as his health permitted, he was a persistent worker.

After many and well-devised experiments Mayow concluded that the air consists of at least two portions, one that supports combustion and sustains life, and another part that does neither. The former he called "fire-air," because it keeps a flame going. He also said that it consists of "nitro-aerial particles," that is, particles in a gaseous form of the kind that makes a mixture of niter and charcoal, or other combustible, burn, when lighted, in the absence of air— even under water.

All this is true enough, but his proof that the air consists in part of a special constituent, different from all the rest, was not complete. He did not collect "fire-air," the gas we now call oxygen, in a practically pure form and show that it is identical with the "fire-air" of the atmosphere. However, he recognized the incompleteness of some of his arguments, and it seems likely that if he had lived a few years longer his proofs would have been perfected, and our knowledge of the composition of the atmosphere set forward almost a hundred years. But this near attainment to the goal, and also even the direct route to it, appear to have been lost sight of for nearly a century. Indeed "fireair," that came so near to being the first constituent of the atmosphere to be discovered, except, of course, water-vapour, and whose properties make it the most conspicuous, turned out to be the very last of all the major ones, save only argon.

Carbon Dioxide

The first of the permanent gases of the air of nearly constant quantity to be clearly discovered was not, as would seem most likely, either oxygen or nitrogen that together make up nearly the whole, but carbon dioxide, that is present as scarcely more than a trace—three parts in ten thousand. At that time alkalies were given to persons suffering with urinal calculi, and certain physicians recommended lime water for the same purpose. Which the view of finding something still better Black undertook investigations with magnesia alba, a form of carbonate, as we now know, or, more exactly, basic carbonate, of magnesium.

On strongly heating this substance a gas is given off, and Black turned his attention particularly to that gas, or air, as all gases were then called. He tried calcining, or burning to a powder, various substances, such as limestone, that we now know to be carbonates, and studying the gas thus obtained. In the end he found that this gas is much heavier than ordinary air, that it will not support life or combustion, that it will recombine with the calx, or powder, produced by the strong heating of the original substance, and that heat will again expel it as before. This gas seems to be fixed, or somehow fastened, in the objects from which it may be obtained.

Nitrogen

The composition of the air in a closed vessel after it no longer would support combustion or life, and to determine the cause of its unwholesomeness. After burning charcoal, phosphorus, or other combustible, in a closed volume of air, as long as possible, he removed the "fixed air" (carbon dioxide), if any had been formed, by means of lime, or an alkali, all in accordance with the previous investigations of Black, and then examined the remaining gas. He showed that this residue is not ordinary air because it supports neither life nor combustion, and that it is not fixed air for the alkalies do not absorb it. He called this residue "mephitic air," because it does not support life.

We now know and say that this residue had been obtained by burning out the oxygen of the confined air and then absorbing the carbon dioxide thus produced. And we know, too, that this residue, this "mephitic air," was nearly pure nitrogen. But at that time oxygen had not yet been discovered, and of course Rutherford could not talk in terms of things and chemical reactions then unknown. He did know, however, that it was obtained by combustion in an inclosed or limited volume of ordinary air, and therefore concluded, after the philosophy of his day, that it was atmospheric air combined with, or modified by the addition of, phlogiston—a mysterious fire substance whose escape from an object commonly is manifested by flame. Nevertheless, and no matter what his ideas as to its nature, Rutherford did obtain reasonably pure nitrogen and did record some of its properties, and therefore may be regarded as the discoverer of this constituent of the air, the most abundant of all.

Oxygen

Almost immediately after the discovery of nitrogen, the other major constituent of the atmosphere, oxygen, was independently found by Joseph Priestly. Priestley was a preacher by occupation and a chemist for recreation. At one time, while waiting for the building of the parsonage to be finished, he had the good fortune to live next door to a brewery—good fortune, because he was induced thereby to take up the chemistry of gases, his chief avocation for many years, on which he published several volumes, and which made his name famous. In the course of this work he obtained a red powder by heating mercury in the presence of air, and then on more strongly heating this powder with a burning lens he got a gas which supported combustion much better than ordinary air. He had gotten oxygen, as we now know, by first burning mercury to an oxide and then decomposing that oxide by raising it to a high temperature.

This element of the atmosphere he called dephlogisticated air. Combustible things burned in it readily; their phlogiston, or fire principle, rushed into it as air into a vacuum. It was air, that is, a gas, but air deprived of phlogiston. But regardless of what he called it Priestley had discovered a new constituent of the atmosphere, the one we now call oxygen, the one that sustains life and supports combustion. On the other hand, Scheele was a professional chemist, so completely absorbed in his subject that he had little time for anything else, whether occupation or diversion.

His incentive to study the air was his wish to solve the riddle of fire, a thing essential to so many chemical processes. He found substances, such as phosphorus, which on burning reduced the volume of the confined air (at the same temperature and pres sure) in which they were burned by about onefifth. In all cases the remaining air would not support combustion.

He also found many ways of getting a gas that would support combustion far better than ordinary air, that was heavier than the air left after combustion, and that made burnt air indistinguishable from ordinary air when mixed with it in the proportion of one volume to four, or thereabouts. The residual or burnt air, the nitrogen, essentially, as we now name it, he called "vitiated air." The other portion, the part that supports combustion, he called "fire-air," just as Mayow had called it a century before.

Scheele, like Priestley and every other chemist of his day, except the immortal and tragic Lavoisier, interpreted fire phenomena in terms of phlogiston. He therefore considered "fire-air," which we call oxygen, to be a combination of phlogiston and a subtle acid substance. But whatever his opinions may have been as to its possible composition, he did find this constituent of the atmosphere, and that too even before it was found by Priestley.

Another student of the atmosphere that must be mentioned in connection with the discoveries of its composition was Henry Cavendish. He may not generally be credited with the discovery of any one of these constituents, and yet he appears to have found nitrogen about the same time that Rutherford did, if not earlier, by the simple process of passing the same confined air back and forth over red hot charcoal and then removing the fixed air (carbon dioxide) with an alkali. He did not publish this—he appeared always to be indifferent about publishing his investigations—but described it in a letter to Priestley in 1772, the year Rutherford published his discovery of mephitic air.

Lord Rayleigh had been making careful measurements of the densities of different gases, and thus came upon the fact that the nitrogen of the atmosphere, or rather that residual then regarded as pure nitrogen, was a little denser than the nitrogen obtained from any one of several chemical compounds.

This was capable of three or four different interpretations, one of which, then regarded as the least likely, was the presence in the atmosphere of an unknown gas denser than nitrogen. At that stage Sir William Ramsay joined Lord Rayleigh in a common attack on this problem. Each tried removing from the air the then known constituents, using entirely different methods for getting rid of the nitrogen, and each found conclusive evidence of the presence in the atmosphere of an unknown gas to the extent of about one per cent. It proved to be chemically inert, and for that reason was called argon.

Other Inert Gases

During the years 1895-1898, that is, immediately after the discovery of argon, four additional inert gases, helium, neon, krypton and xenon, were found by Sir William Ramsay and his assistant. Travers, to be constituents of the atmosphere.

Helium was found as that portion of the atmosphere that is still gaseous while the rest has been chilled to liquids or solids; and the others, neon, krypton and xenon by fractional distillations of large quantities of liquid air and liquid argon, impure, of course.

HYDROGEN

Lord Rayleigh and others also found free hydrogen always present in the air, but in amounts that varied from, roughly, one part in 1,000,000 to one 5,000. Perhaps this gas can not be regarded as a permanent component of the atmosphere in the same sense that oxygen and nitrogen are, but rather as an accidental and variable impurity. At any rate it is irregularly added to the air, at least by volcanoes, and more or less irregularly removed from it.

Traces and Impurities

Varying traces of ammonia, nitric and nitrous acids and their compounds, sulphuric and sulphurous acids and their compounds, oxides of nitrogen, hydrogen dioxide, and ozone are among the innumerable things always in the air if not of it. Minute particles of sea salt from evaporating spray, fine earth dust caught up by winds, pollen of every description and spores of many kinds are among the coarser and ever present pollutions of the atmosphere from the earth beneath. Shooting stars furnish a continuous, though invisible, shower of dust the world over from outer space.

Also radioactive products of radium, thorium and other elements are pouring into the atmosphere continuously and contributing to the maintenance of its electrical state.

This state means, in part, around 20,000 electrified particles, or ions, per cubic inch in the lower air. The number of ions 60 miles or so above the surface is far greater, at least 1,600,000 per cubic inch, as we infer from the phenomena of radio communication.

In short, while the things of the atmosphere are but few the things in it are innumerable. All these latter are relatively very small in amo t, but some of them are exceedingly important. The salt particles and some others are essential to condensation and rainfall; the ammonia and other nitrogen compounds, brought down by precipitation, add much to the fertility of the soil; the ions of the high atmosphere make distant radio communication possible; and ozone, also in the upper air, shields us from that portion of the ultra violet radiation that would destroy our eyesight, as we are now constituted, and otherwise do us irreparable harm.

PERCENTAGE COMPOSITION OF PURE DRY AIR

If the atmosphere is freed from all its numerous impurities and also freed from water vapour, there will remain pure dry air whose percentage composition at the surface of the earth is, very closely:

The amount of water vapour in the atmosphere, or better, perhaps, the amount of the water vapour constituent of the atmosphere, varies from scarcely more than a trace, at extremely low temperatures, to at least 5 per cent by volume on the hottest and most humid days. At and above the height of 5 miles, say, the amount of water vapour always is small, even when saturation obtains, owning to the very low temperatures at these levels. ts average value, the world over, is such that if it were all condensed it would be the equivalent of a layer of water about one inch deep over the entire earth.

Nitrogen	78.03
Oxygen	20.99
Argon	0.9323
Carbon dioxide	0.03
Hydrogen	0.01
Neon	0.0018
Helium	0.0005
Krypton	0.0001
Xenon	0.000009

FORMS OF OXYGEN AND THEIR DISTRIBUTION

Oxygen also is peculiar in its distribution, and it occurs in three different forms. All but about one part in 400,000 is ordinary oxygen, or, in the language of the chemist, diatomic oxygen. Triatomic oxygen, or ozone, occurs chiefly beyond the highest clouds, its greatest density being, apparently, at the levels of 20 to 30 miles. It occurs in the lower air only as a trace. In the highest air, 60 miles and more above the surface, the oxygen appears to be at least partly monatomic, according to the spectrum of the aurora.

Importance of Ozone

Every one knows that life would be impossible if there were no ordinary or diatomic oxygen in the atmosphere, and that without it nearly if not quite all vegetable life also would be impossible. But it is not so well known, though equally true, that animals, including the human species, could not exist as now constituted if the air did not contain a small amount of triatomic oxygen or ozone.

And yet, paradoxical as it may if seem, ozone were just a little more effective in its goodness, again life, as now constituted, could not last. These surprising facts come about in this way: The radiation from the sun includes not only every colour, that is, the whole of the visible spectrum, but also extends indefinitely beyond the red into the long-wave-length invisible region, and likewise, in the opposite direction, well beyond the limit of the violet.

But in this ultra violet portion of the spectrum the radiation from the sun ceases far short of the limit to which that from an electric arc, for instance, can be followed; and ceases, not because no radiation beyond that limit is given out by the sun, but because it is absorbed by the ozone in the upper atmosphere. Now, much of that particular radiation which ozone absorbs is destructive to the eye, and when intense probably injurious to other tissues as well.

This particular limiting, then, of the solar radiation that reaches the earth is amazingly located. If it was a little farther out in the ultra violet, eyes, as now constituted, could not have developed; and if a little nearer the visible, again; animals, as they now are, could not have come into being. Of course ozone, in moderate amount, presumably was in the air long before there were animals of any kind to be affected by its action on solar radiation. This radiation, therefore, was not adapted to them, but they developed in adaptation to it; nevertheless the fit is so close and of such a surprising nature as to give us decided pause for thought.

Origin of Ozone

Ozone is produced by the action of extreme ultra violet light on oxygen, and therefore at great heights; a little where the oxygen is very rare, then more and more with decrease of height and increase of the oxygen supply until by absorption the ozonizing rays are considerably enfeebled. As the density of the oxygen increases the intensity of the effective radiation decreases, hence the rate of production of ozone by this process must be very slow in the outermost air, increase to a maximum with decrease of height and then rapidly fall off at still lower and lower levels.

The total range of ozone, top to bottom, probably is not less than 100 miles, with its maximum concentration around 25 miles, perhaps, above the surface, and yet all told, and despite its great importance, it is the equivalent of a layer of this gas only about one-tenth of an inch thick at atmospheric pressure and room temperature. Ozone is produced also by electric discharges through oxygen. Lightning certainly produces some ozone, and it may be that the auroras form it too, though the great heights at which they occur, 60 miles and more above the earth, seem to render this conclusion doubtful.

Distribution of Ozone

As already stated, we know that at most there is only a trace of ozone in the atmosphere up to the level of the highest clouds, and that it exists to an appreciable extent at considerably greater heights. But this is not all. The total amount of ozone vertically over one to the limit of the air appears to increase, in general, with increase of latitude; to be greater during winter and spring than summer and fall; and to be greater in winds from high latitudes than in those of the opposite direction. This complicates the problem of the origin of ozone. If it is produced by ultra violet radiation alone why should it not be most abundant in tropical regions, and elsewhere in the summer time? If we surmise that it is produced largely by auroral discharges, how, we ask, can these discharges, 60 miles up and more, reach a sufficient supply of oxygen? And if there is enough oxygen at these great heights how can the ozone subsequently become concentrated at the intermediate levels? These are some of the questions to be answered by future observations and studies.

UPPER AND LOWER ATMOSPHERE

The composition of the lower atmosphere up to at least 6 or 7 miles in temperate regions, and 8 or 9 within the Tropics, is well known, except in respect to condensation nuclei and certain impurities.

Throughout this region too the percentages of the several gases, except water vapour, are practically constant owing to their continual mixing incident to convection and turbulence. Somewhere in the upper air, however, the percentages of the lighter gases must increase and those of the heavier decrease under the action of gravity since at these levels vertical convection is very feeble.

But as the upper air grows thinner the more and more rapidly does our knowledge of it become less. We know that its outermost portion is the very extensive region of the aurora, hundreds of miles thick; that near the base of the auroral region there is enough ionization to make wireless communication around the world entirely practicable; and that far below this Kennelly- Heaviside layer, in turn, and yet well within the upper air occurs most of the ozone, the triatomic oxygen that indirectly is so vital to all terrestrial life. We seldom give any thought to this upper air, but it is so important that we really must know more about it.

Mass of the Atmosphere and of its Constituents

From the known percentages of the several constituents of dry air, given above, their molecular weights, and various other pertinent facts such as the amount of water vapour present, height of the barometer, volume of land above sea level, and distribution of temperature with height, it is easy to compute the approximate mass of the atmosphere as a whole and of each of its several gases.

The results are given in the following table, in which the factor 108 means: Add eight ciphers, or multiply by 100,000,000. Since the values in this table were determined more or less independently it could not be expected that the percentages found of the constituents would add up to exactly 100, nor that the sum of the computed masses of the several parts would precisely equal the mass of the whole. These deviations, however, are very small—probably within the present limits of experimental errors.

Table. Mass of the Atmosphere and of its Constituents in Tons (2000 Pounds)

Substance	*Volume per cent dry air, at surface*	*Total mass*
Total atmosphere		56,328,000 × 108 tons
Dry air	100.00	56,181,850 " " "
Nitrogen	78.03	42,684,725 " " "
Oxygen	20.99	12,782,647 " " "
Argon	0.9323	682,125 " " "
Water vapor		146,150 " " "
Carbon dioxide	0.03	23,874 " " "
Hydrogen	0.01	1,423 " " "
Neon	0.0018	759 " " "
Krypton	0.0001	141 " " "
Helium	0.0005	88 " " "
Ozone	0.00006	33 " " "
Xenon	0.000009	19 " " "

The numbers here given that express the masses of the atmosphere and its several constituents are useful as quantitative values and for exact comparisons, but so great, even though in terms of tons, that we can form no distinct conceptions of them—they are just awfully big! A clearer idea may be gotten from the fact that the total mass of the atmosphere is the equivalent, roughly, of that of a block of granite a thousand miles long, a thousand miles broad and half mile thick; while the least abundant of the constituent, xenon, if loaded on cars, 19 tons to the car, would freight a train reaching 40 times around the earth along a great circle, and which, traveling 20 miles all hour, would be 6 years in passing any point on the road.

❒

2

Forms of Condensation of Precipitation

The meteorologist borrowed the term precipitation from the chemist and then generally gave it a meaning different from that used by its original proprietor. The chemist means by it the process of separating a substance from its solution in such form that it settles or tends to settle under the force of gravity. The substance itself he calls a precipitate. The meteorologist sometimes makes this distinction too, but more commonly he means by precipitation not a process but a product, such as rain or snow.

The chemist gets nearly all his precipitates from solutions, and commonly, too, as a result of some combination into a new substance. The precipitation (precipitate) of the meteorologist, on the contrary, is not a new substance it all, but the same old substance in a different state, that is, liquid water, such as rain or solid water, as snow, for instance, condensed from gaseous water. Some of the chemist's precipitates settle quickly, while others fall slowly, and a few take forever-and-a-day to come down. So it is also with the meteorologist's precipitations. Large hailstones fall with a speed that is dangerous, drizzle drops just dawdle along, and dew doesn't fall at all—it is just said to fall.

There is a tendency on the part of many meteorologists to confine the term precipitation to only that water, liquid or solid, which, after dropping out of the clouds, actually reaches the earth. In its broader usage, however, it includes, as explained, all natural condensates from gaseous water to liquid or solid water. It is here used in that more comprehensive sense.

FOG

One of the most interesting, and in some respects an exceedingly important, form of precipitation, is fog, a great swarm-like assemblage in the surface air of hundreds of thousands of droplets per cubic inch so minute that it would take 7,000,000,000 of them to make a teaspoonful of water. It is a surface cloud whose myriad myriad droplets have formed on condensation nuclei, such as ultra-microscopic particles of seasalt, incident to the cooling of the air below its dewpoint. In many cases this necessary cooling occurs on still clear nights in river and creek valleys, as the result, chiefly, of the loss of heat by the surface through radiation. When the sky is overcast with low clouds their radiation to the earth is nearly equal to that of the earth to them, and the cooling of the surface therefore, and, through it, the adjacent air, insufficient to induce condensation.

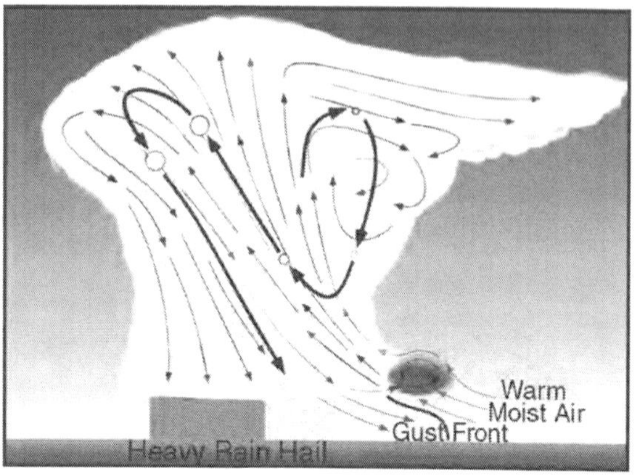

Fig. 2.1: Radiation fog

Similarly this type of fog, radiation fog as it commonly is called, can not and does not form, even on clear nights, when there is a considerable wind, as the cooling then also is slight. The loss of beat by radiation is not, of course, affected by wind, except indirectly through its effect on the surface temperature; but the fall in temperature and the depth of the chilled air, and hence the formation of fog, are affected by it. Thus, whenever a dead calm prevails the cooling surface of the earth cools in turn only a very shallow layer of air, practically just that which is in direct contact with the chilled objects on the surface, since it itself is a very poor conductor of heat. This chilling of the mere surface air, no matter how pronounced, does not produce fog, a sort of cloud above the surface, but only dew or frost condensed from the air that is chilled by the cold objects with which it is in direct contact.

If there is a little wind, just enough to mix the air through a layer 10 to 50 feet deep, the cooling, though less than when there is no wind at all, often is enough to push the temperature of this mixed air below its dewpoint and thus to fill it with fog. That is, the considerably cooled surface air, by mixture with a rather small amount of air above, cools it in turn below its point of saturation and thus induces throughout its volume condensation on the ever present, innumerable, hygroscopic nuclei, in droplets too small to fall to the ground against the diffusing force of the existing slight turbulence caused by the slow wind movement.

When the winds are strong, however, the consequent pronounced turbulence brings, one after another, all portions of the air through a considerable depth into contact with the surface and thus the loss of heat, though actually greater than it is on still nights, owing to higher surface temperatures, now is shared by a far larger mass of air and the fall in temperature, or actual cooling, thereby much reduced, and the occurrence of fog prevented no matter how bright the stars. In short, even when the skies are clear, no wind at all implies no fog; very light wind, say 2 to 6 or possibly 8 miles per hour, fog, if other conditions are favourable; strong wind, again no fog.

Fog is formed also by the drifting of warm humid air onto a cold surface off Newfoundland where the air from the Gulf has come over the much colder Labrador

Current. It forms also, though to a much less extent, where the air is distinctly colder than the water. This is an exceedingly common occurrence on calm cold mornings over ponds, lakes and streams, from which one often sees wisps of fog rising at numberless places like so much steam from hot water.

That is why the usual name for this phe- nomenon is steaming. It is only the condensation of the moisture in the humid surface air as it is buoyed up and mingled with the colder air above. Ordinarily the evaporation is so slow, and the air above so dry, that this steam-fog is only a shallow, gauzy affair of but little consequence. Occasionally, a in high latitudes, the air is so very cold that the "steam" immediately changes to ice needles, or frost smoke, as it usually is called.

Fig. 2.2: Advection Fog

Not only may fogs be produced by the drifting of warm humid air over cold surfaces, and by the importation of cold air over warm water, but also, however and wherever formed, they themselves may be transported by gently moving air to other places, even in many cases to localities at which fog seldom is generated in situ. Such advection, or drift, fogs are exceptionally frequent just inland from mid-latitude seacoasts along which the prevailing wind is on-shore.

In certain cases this moving fog is largely caught by trees, notably by the towering redwoods of California, as it filters through them, and thereby condensed into a fog-drip that in every particular is the equivalent of just so much gentle rain. And this rain is so frequent too in various places as to maintain luxurious vegetation where otherwise only the spiny, stunted growths of an arid region could survive.

CLOUD

In general, a cloud, a thing that every one knows when he sees it, though few know much about it, is essentially just a fog in the sky, though on the average its droplets are roughly 35 times as bulky as those of true fog—so bulky indeed that if they were merged together a teaspoon could hold only a paltry 200,000,000 of them without running over. But not all clouds consist either wholly or even in part of these droplets. Some contain also rain drops, each of which is equal in mass to many thousands of the average cloud particle, while many others contain no droplets at all, but only snow in some one or more of its many forms, from discrete, tiny columns or needles to matted tufts an inch or more across.

Clouds differ from each other not only in respect to composition, some consisting of water droplets and others of ice particles, as just stated, but also in height above the surface, appearance and mode of formation. And these conditions are the criteria according to which we classify clouds and name them—a delightful intellectual diversion for which our forefathers felt no need and never practiced.

When they had occasion to speak of a cloud they generally just called it a cloud, and let it go at that, without specifying what sort of a cloud it was. In fact it was not until 1801 that clouds were christened at all in any European language, and until 1803 that they were given names that proved to be generally acceptable. A second edition of this international cloud atlas, only slightly modified and extended, was published in 1910.

In general this atlas was quite satisfactory and became widely known and extensively quoted in books and periodicals. In 1922 another International Cloud Commission got busy and at the end of ten years produced a handsome work that has much to recommend it, especially its many excellent cloud pictures.

But shade of Luke Howard! How some of the good old definitions have been changed, and how unwieldy for practical use, are both it and the abridged form that appeared in 1930! Many of us expect another International Cloud Atlas to appear some day in which the revisions largely will be reversions. Also there is urgently needed a working cloud atlas, cheap and above all convenient to use.

Fig. 2.3: Cirrus, hook type.

All the more commonly recognized clouds are listed below, though the definitions and comments are less formal than those given in international atlases.

Cirrus

The highest, generally, of all clouds, with the exception of two rare forms, to be mentioned later, the natures of which are unknown, is called "cirrus," and was so named by Luke Howard, because of its frequent appearance in the form of swirls and ringlets. It is a detached, white cloud, fibrous in structure and too thin to cast a distinct shadow or blur the outline of the sun or moon.

Fig. 2.4: Tufted Cirrus

Owing to the great heights at which it usually occurs, 6 to 7 miles above sea level in mid-latitudes and half as high again near the equator, its temperature commonly is very low, -50° to -80°F roughly. And because of this very low temperat-ure the citrus cloud nearly always, though perhaps not quite invariably, consists of ice particles.Although the cirrus usually is higher than any other cloud, it does not always have that topmost position. Indeed it can and does occur at any level within the convectional portion of the atmosphere-the troposphere—from its highest reaches clear down to the surface of the earth. It occurs now and then beneath a cloud sheet of a different type and even between two such sheets or layers.

Furthermore, it assumes as many different forms as Proteus, to many of which distinctive names, seldom used, have been given. When it slowly disappears it is a harbinger of fair weather, but when it grows denser it frequently is the forerunner of a general rain—a sort of a pilot or scout for the ranks of heavier and lower clouds following in its wake.

Fig. 2.5: Cirro-stratus.

Cirro-Stratus

The next highest cloud also is white and commonly too thin to blur the outline of the sun. It may appear as only a formless milky haze or as a more or less continuous, fibrous sheet or layer.

It therefore has been given the appropraite name "Cirro-Stratus," that is, a stratum of cirrus cloud. It shows the well-known large rings or halos around the sun and the moon, from which we know definitely that it consists of minute ice crystals, as they alone so refract the light of the sun or moon as to produce these circles. When this cloud tends to thicken there generally are still denser clouds on the way bringing rain or snow in 6 to 24 hours.

Cirro-Cumulus

Sometimes the cirrus, or cirrostratus, seems to curdle, as the housewife says of milk when it coagulates into lumps. Such a cloud canopy, consisting of small lumpy clouds, often with more or less fibrous edges, but practically without shadow, and irregularly distributed, frequently is called a "curdled sky."

Usually, however, the cloud lumps are arranged in orderly rows, or waves, that remind one of the colour patterns on a mackerel's back. This is the "mackerel sky."

Fig. 2.6: Cirro-cumulus.

Not infrequently, too, the larger waves are crossed, diagonally, as a rule, by many much smaller waves, or ripples. Technically all such fine lumpy clouds are called cirro-cumuli, or heaped-up cirrus. They obviously are caused by numerous small convections in a thin unstable layer of air, and evidently contain but little cloud material since they are practically without shadow.

Alto-stratus

The next cloud type in the order of decreasing altitude is the alto-stratus, a heavy, whitish to bluish veil of clouds, sometimes fibrous in texture and again structureless, which we call alto-stratus. Commonly it is thin enough to reveal the approximate position of the sun, as if seen through ground glass, but too thick to show either coronas or halos.

Fig. 2.7: Alto-stratus

It may be formed in any one of several ways, chiefly by the spreading out by wind of the upper portions of cumulus clouds; by the cooling in place of a layer of moist air; and by the flow of a layer of warm humid air over colder air, the cloud forming along the interface. It is not disturbed by numerous local convections as we know from the fact that they would make of it an altocumulus, a cloud it does not resemble; neither is it a product of a wide-spread active convection, since that would cause a more or less steady rain or snowfall, which is not a feature of the alto-stratus. It is recognized that the above definition of alto-stratus is in keeping with that given by the older international cloud committee, appointed in 1891, and not with that of the committee of 1922. According to this later definition the altostratus always is fibrous, at least in parts. If so, then the fairly common "ground-glass" cloud of the same level and density that shows no such structure is without recognition. In this case many of us prefer our old friend to the new acquaintance.

Alto-cumulus

Somewhat lower, on the average, than the alto-stratus is the alto-cumulus, a canopy of closely packed but isolated or nearly isolated cloud masses, seldom more than a few diameters of the sun or moon across and nearly always arranged in more or less conspicuous rows, often in two directions. The smaller of these cloud masses are white; the larger appreciably shaded. When these clouds are slowly evaporating, and are 20° to 30° from the sun, they often afford wonderful colour displays—especially the delicate pinks and greens that characterize the finest mother-of-pearl. These exquisite patches of colour are but fragments of coronas produced by the diffraction of sunlight by water droplets. This iridescence in addition to charming us with its beauty, tells us also that the alto-cumulus consists of water droplets and not of ice crystals. Droplets produce coronas, iridescence, and rainbows; the crystals give the many forms of the halo.

Fig. 2.8: Alto-cumulus.

The alto-cumulus, like the cirro-cumulus, appears to be caused by local convections in an unstable layer of air, but one of greater moisture content than that in which the cirrocumulus forms. When the quantity of the humidity present is unusually great an isolated cloudlet here and there grows larger and much taller than its neighbours. In this case the cloud sheet may be called alto-cumulus castellatus, and is so called when are especially discriminating. Such a cloud not infrequently is a forerunner of rain. In practice, it is the whole cloud sheet that is called altocumulus, and not one of its individual parts to which, ordinarily, no name at all is given, though actually it is a cumulus.

Strato-cumulus

There is yet another cloud sheet of irregular density, a sheet of largish, dark grey balls or rolls of cloud that frequently are united by thinner cloud. These masses are larger, denser and commonly at a lower level than those of the alto-cumulus. Also in in many cases, especially just after the passage of a general storm, this cloud has the form of a continuous canopy in irregular folds and layers. Another type of this cloud is low and fluffy in appearance, and often, presumably, produced by the convection due to turbulence.

The tops of the unit masses of this cloud often are raised and rounded, showing that they are produced each by a local convection, while their bases have an approximately common level. In many cases, however, the tops of the separate masses spread out, like smoke under a ceiling, until they nearly or quite merge together. And they spread out because they too, like smoke in a room, have reached a ceiling—the limit to which the distribution of the temperature in the atmosphere permits them to rise. Just how and why this is it would be interesting and instructive to follow up in detail, but to do so soon would have us repeating in part the important story of the structure of the atmosphere and repetition to such an extent is not in order.

Nimbus

When a general, or cyclonic, storm is passing the clouds associated with it often run the changes from cirrus, as the most advanced scouts, to cirrostratus, and then an altostratus that grows denser, darker, and lower, covering the entire heavens and without distinctive form, until finally rain or snow falls from it steadily. In this final state, and even before that, so long as rain or snow seemed imminent, this cloud sheet formerly was called nimbus, and that is what some of us still want to call it who see no need whatever for giving it the new name "nimbostratus" adopted by the latest cloud committee. The two names refer to one and the same cloud, and therefore the new and longer name can only take more time to write, more space to print and cause more cloud confusion, a confusion that already was so great as to shake ones confidence in many, if not most, cloud observations.

The nimbus cloud normally results from the gradual ascent of humid air over an obstacle in its path, rising land in some cases but more commonly cooler air in front of the storm centre. Some rain or snow can and does fall from nearly, if not quite, every type of cloud, and even, on rare occasions, out of a clear and cloudless sky but that does not justify calling anything nimbus except the cloud just described, from which rain or snow normally falls in appreciable amounts.

Fig. 2.9: Strato-cumulus.

Cumulus

When the weather is calm and warm and the sky generally clear, surface heating by sunshine starts rising many a column of air here, there and yonder, and when that air is moderately humid a cloud forms at the top and in the upper portion of each well-developed pillar—a cloud rounded above, with bulging sides, like a great pack of wool or a giant head of cauliflower, and flat at the base. This is the cumulus cloud, brilliantly white where the sun shines on it, and blue-grey to dark over the shaded portion, except along the edge not far from the sun where there is a silver lining—a wondrous capital of marvelous carving supported by an invisible column. Some of these clouds tower higher than their neighbours, but the bases of all, big and little, at any particular time and place, have substantially the same height, the height at which the current surface air on rising will have cooled to its dewpoint, or temperature at which saturation is attained and condensation begins.

Fig. 2.10: Cumulus.

During its growth the air within a cumulus cloud is in a state of vigourous agitation, as is evident from the way its sides and top continuously change in size and shape, and confirmed by the jolts and tossings the aviator experiences when in a large cloud of this type.

From the origin of the cumulus cloud, namely, vigourous local convection, incident to surface heating, of moderately humid air, one correctly infers that in middle latitudes it is essentially a summer phenomenon; that it is most numerous in tropical regions and rarest near the poles; that at times it forms a chain of balloons, as it were, in the sky that delimits the course of a river, or a seashore, below; that over land it is most numerous by day and over the ocean at night; that it frequents islands far more than the open ocean; and that, among many other things, it is a fair weather phenomenon, since it can not occur when the sky is generally overcast, as adequate surface heating does not then occur, nor at the time of any considerable wind, since that would prevent the establishment of large columns of ascending air.

Cumulus-nimbus

When a cumulus cloud grows to great size and height by the continuous feeding into it from below of humid air, or has been formed along a squall line by the juxtaposition of warm and cold air, the time comes when for one reason or another vast numbers of the droplets become so large and heavy that they fall out as so much rain which commonly reaches the ground, though in arid regions it sometimes evaporates wholly in mid air. As soon as this rain phase of the cumulus cloud begins we heretofore have been ready to change its name to cumulo-nimbus.

As the rain begins thunder and lightning also usually occur. In many cases also the cumulo-nimbus spreads out at the top into a more or less welldefined anvil shape. Finally, the topmost portion of the cloud frequently, but by no means always, nor at first, is combed out by the wind into a sheet of cirrus. And yet this is the one phenomenon by which the 1922 Cloud Commission distinguish between the cumulus and the cumulonimbus. They say: "Masses of cumulus however heavy they may be, and however great their vertical development, should never be classed as cumulo-nimbus unless the whole or a part of their tops is transformed into a cirrus mass."

Again: "None of these cumulus clouds should show ice crystal clouds (hybrid cirrus) at their tops; this would mean that they had reached the cumulonimbus stage." And in the formal definition of cumulo-nimbus there is this phrase: " the upper parts having a fibrous texture." Finally, their brief definition of cumulo-nimbus is: "Cumulus clouds of great vertical development, with the tops composed of ice crystal clouds."

Well, if we must accept this definition of cumulo-nimbus, then this is the definition of cumulo-nimbus we must accept, being both official and international. But, for all that, it leaves some of us a little confused in our minds, as we say-those of us who have seen many a heaped-up cloud from which abundant rain was falling and in which there was frequent and vivid lightning, but of which no portion was cirrus nor showed any signs of cirrus.

Fig. 2.11: Cumulo-nimbus.

We called such a cloud cumulonimbus, and we very much dislike now to call it just cumulus. Indeed, rain is the really important differentiating feature between cumulus clouds, and not the trifling circumstance of the presence or absence in their uppermost portions of a little cirrus. And our confusion is greater, not less, when we recall other heaped-up clouds, cumuli we then called them, and still would like to call them, which developed abundant cirrus, but never furnished a drop of rain or a flash of lightning.

This is another case where some of us much prefer our old and faithful friend to the new and untried acquaintance.

STRATUS

Sometimes the sky is covered with a comparatively low layer of grey formless cloud from which little or no rain falls, only a drizzle at most. This cloud is called "stratus," and the height of its base varies from just free of the surface to about 3000 feet. It may be formed in several ways, such as the flow of warm humid air over colder air beneath, the cloud forming at their interface; by the drift of fog over relatively warm land, where the lower portion of the fog is burned off, leaving the upper portion as a stratus cloud; and by the relatively shallow turbulence convection induced by surface friction and surface obstacles. Stratus is the lowest and most nearly uniform of all clouds. Indeed it is so devoid of contrasts that it scarcely can be shown in a photograph at all except where it is pierced by a hill or mountain, and even in that case a lively imagination is a great help—not to supply the details, for of these there are none, but to awaken that lonesome, dreary feeling one has on a day of drizzle. It would seem that, in all conscience, Nature had completely exhausted her sartorial ingenuity in the making from vapour alone of all the above sorts and kinds of cloud robes. But here too her ability has no bounds. There were scraps left over, as always, when the robes were fashioned.

BILLOW CLOUD

Although the billow cloud is only a special form of the alto-cumulus, it is so strikingly different from the average cloud as to arrest attention and to call for a distinctive name. It occurs in long, narrow and evenly spaced strips of white cloud. Even the individual strips themselves often show considerable detail of structure. It generally is supposed that this cloud occurs on the crests of waves or billows formed by the flowing of one current of air over another.

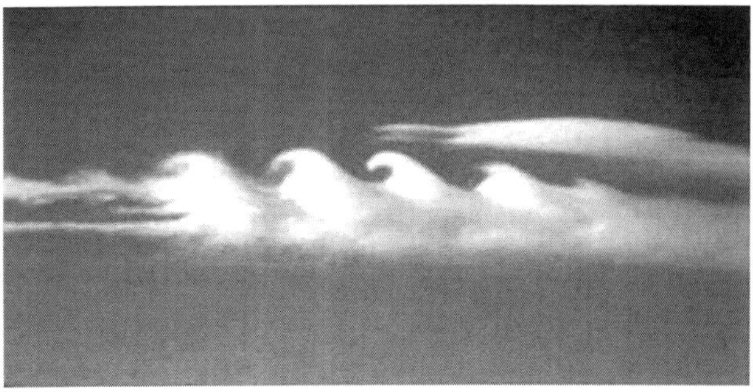

Fig. 2.12: Billow clouds.

However, while the name "billow cloud" generally is restricted to a particular form of the alto-cumulus, the same sort of structure though of much finer grain, occurs as a frequent type of the cirro-cumulus, and likewise a far coarser one as occasional variety of the strato-cumulus.

As just implied, two sheets of air of different density flowing horizontally the one over the other develop waves or billows at their interface which billows of course are coolest in their highest portions or crests and warmest in the troughs. Hence when the humidity is just right such billows are cloud-crested and clear-troughed. But according to the theory of these air waves the distance from the crest of any one to that of the next is far greater than that from cloud to cloud in the delicate cirrocumuli ripples, and greater too than the distance between the rows of a wave-like alto-cumulus or billow cloud.

Hence, while actual air waves, induced in the manner explained, probably are the cause of certain groups of cloud strips of coarse grain, or long wave length, it seems probable that the beautiful alto-cumulus billows are not produced in this manner, and certain that the citrus ripples have some other origin. The probable origin of many, perhaps all, of the finer structured, or shorter wave-length billow clouds, is convection through unstable air beneath into air above of different velocity.

Fig. 2.13: Lenticular cloud.

When the two superincumbent layers of air are still with reference to each other the ascending masses are roughly as broad as long, horizontally, as indicated by the roundish patches of cloud that often cap them, and usually occur in more or less distinct rows and patterns. When, however, they are flowing the one past the other the clouds are more or less drawn out in the direction of the motion of the upper air with reference to the lower, and if this motion is in the direction of an alignment of the clouds the individual small cumuli in each row parallel to the wind are so intermingled, or so it seems, as to become a uniform cloud wave or billow. This obviously is not the whole story of the genesis of the billow cloud, and frankly it may not be more than half right as far as it goes. At any rate this cloud still challenges us for much more observation, experiment and study than yet has been given to it, as do also cirro-cumuli waves, especially those that are crossed by numerous ripples.

Lenticular Cloud

Sometimes an isolated cloud, whether solitary or one of a group, has the general shape of a double convex lens, that is, thick in the middle and gradually thinning to nothing at the edge. This is the lenticular cloud, whatever else it may be, such as alto-cumulus, alto-stratus or what not.

It may be formed by the arrest of an ascending small cumulus cloud at the base of a layer of stable air—a layer in which the decrease of temperature with increase of height is so slow that ascending air, unless in great volume, can not deeply penetrate it. Here the cloud, arrested by the impenetrable ceiling, spreads out much in the shape of a lens, circular or distorted as the circumstances, especially the direction and strength of the winds, may determine. Another abundant source of lenticular clouds is any large mountain peak, ridge or range. Here, in most cases, the crossing wind is thrown into great standing billows whose crests occasionally are capped with lenticular clouds as fixed in position as a waterfall, and made of as rapidly changing material—condensing on the windward side and evaporating to the leeward.Sometimes this lenticular cloud of obstruction or interference origin, and therefore cresting a stationary air wave, is low down, and again at considerable height, as determined largely by the amount and distribution of the humidity. This cloud also deserves more study than it yet has received.

Crest Cloud

When quite humid wind blows up and over a mountain a cloud forms, such as the famous "TableCloth" Cloth" on Table Mountain near Cape Town, that rests on the ridge and hangs down each side like a giant cloth, fringed on the leeward slope where it is pulled down in evaporating shreds by the descending air. This is the crest cloud. It is interesting and of some importance, and if baptized in Latin might eventually become a member of the aristocratic cloud family.

Fig. 2.14: Lenticular Cloud, Interference Type

Riffle Cloud

Frequently, when the crest cloud is moderately heavy and the wind fairly strong one or more long cumulus billows hang fixed in the free air to the leeward of the mountain ridge and at about its height.

Fig. 2.15: Crest cloud.

These mark the upper portions of the great air waves caused by the interference of the mountain with the passing wind. Analogously produced waves, called riffles, occur in running water where it passes over an obstruction in the bed of the stream; hence the name "riffle cloud."

Banner Cloud

Any high mountain peak that steaks up like Cleopatra's Needle, and there are such—Mount Assiniboine in Canada, for instance—now then develops a cloud, in an otherwise clear atmosphere, that flutters out like a huge banner made fast to the peak's upper leeward side. This occurs, of course, only in air that is humid well-nigh to the point of saturation. The cooling necessary to pass that critical cloud-producing temperature obviously may be owing in part to contact of the air with the cold walls of the peak, and certainly in part to the decrease of pressure in the eddy region on the sheltered side. It is not a familiar cloud nor one of much importance, but for all that it is both beautiful and interesting.

Boa Cloud

Fig. 2.16: Banner cloud.

Sometimes a great river of cloud, such as may flow along a broad valley, is obstructed by an isolated mountain, a volcanic cone, perhaps, that stands in the midst of the stream and divides it right and left as would an island. Thus the mountain becomes wrapt around by the flowing cloud, as by a giant feather boa. The old Romans called it, when it twined about Mount Etna, the Serpe, and rightly regarded it as a harbinger of rain.

Scarf Cloud

If one will watch carefully just above the top of a large growing cumulus he sometimes will have the pleasure of seeing a little cloud of silken sheen and bowed upward, over and free from the topmost head like a portion of a halo, emerge from the sky where before there was nothing to be seen. In a little while it probably will cover the head of the cumulus, and a bit later rest on its shoulders like a great silken scarf. Some say that it is a harbinger of showers. Perhaps so; at any rate the towering cumulus that generates it certainly is.

MAMMATO-CUMULUS

If you ever saw the sky covered with huge grapes, or everywhere festooned, or a great mother cloud, as it were, plethoric with a thousand mammae, then, whether you knew it or not, you were looking at the interesting mammatocumulus cloud, a frequent forerunner of the dread tornado. Not everything about the genesis of this cloud is known, but it appears to be the result of an upsidedown convection of some sort, perhaps induced by the flow of a thin cloudy layer of air over other air beneath that is sufficiently warm near its upper limit to permit of short-range interchanges between the two.

Tornado Cloud

The tornado cloud is a violently rotating, hollow, tapering cloud column 100 to 1000 feet in diameter, as a rule, extending from the surface of the earth, where it is smallest, up into the heavy storm clouds overhead. The cooling that produces this cloud is caused by the expansion of the air incident to the decrease of pressure within the column due to its rapid rotation. The cloud column of the waterspout is produced in the same way. This is a thin cirrostratus- like cloud of rare occurrence and unknown origin, richly iridescent and about three times as high as ordinary cirrus, that is, around 16 miles.

Fig. 2.17: Tornado Cloud.

Noctilucent Cloud

It was stated in the description of the cirrus cloud that usually it is the highest of all save two that are rarely seen. The higher of these, in turn, is the noctilucent cloud, a thin cloud-like patch of something, roughly 50 miles above the surface. Not all the clouds of this great height are lucent, but the first ones observed were, and the name "noctilucent" then given to them may as well be applied alike to all until we know something—now we know nothing-about the substance of which they are composed and how they are formed

❐

3

Water Vapor Distribution and Trends

It long has been recognized that the first thing to do in cooking a rabbit is to catch the rabbit. Similarly the distribution of water vapour requires that first there shall be water vapour that may be distributed. And this raises questions about that water vapour. Where does it come from? How much is there of it? Where does it go to? How does it get there? Well, no matter what was the origin of the first water vapour about the earth, nearly all of it that at any time is now in the atmosphere got there as a result of evaporation, mainly from water surfaces, snow and ice sheets, damp soil and growing vegetation.

EVAPORATION

In picturing to ourselves the mechanism of evaporation, it is convenient in respect both to gaseous and liquid water (ice presently will be mentioned separately) to think of the ultimate particles or molecules as moving about among themselves along exceedingly short paths (except only when the vapour and all its accompanying gases are together extremely rare) in every which direction and with widely varying speeds with reference, in the case of each particle, to some point in the direction of its path and which we regard as fixed in position. This may sound a little finical, but it isn't, for really we must have something to tie to, as we say. If we measure the velocity of anything it must be in terms of its rate of approach towards, or departure from, something else that for our particular purpose we can consider as at rest, whatever the motion of the thing "at rest" may be in respect to a third object, or of that in turn to still another, and so on indefinitely. For instance, one may consider his face as being at rest while his hands are in motion with reference to it as he eats in a dining car; or the car at rest and himself, face and all, in motion as he comes in or goes out; or the ground at rest while the car is in motion over it; or the sun at rest while the earth has several kinds of motions; or the stars at rest while—but there is no end to this and we would better stop while stopping is good.

The point is this: If we talk about velocity we always must think of, or measure, that velocity in terms of rate of movement directly towards, or directly away from, something else, which for our then particular purpose we may regard as being fixed in position. If this little digression has helped us to think straight about "velocity," so that we know what we mean, we then may return to our consideration of the linear movements (they also spin) of water molecules. In the case of an open vessel of water, for example, the sides and bottom evidently may be regarded as at rest with respect to the moving molecules, and so too may the surface of the water, if it is not agitated. The same also is true of reservoirs, ponds, lakes and other bodies of water.

As just explained, we think of the molecules of water, and also of those of every other sort of liquid or gas, as being continuously in a state of utterly confused migration both as to direction and speed. Furthermore, the greater this average speed the higher the temperature. This is the right way to look at it, but if we wish to do so, and it often is convenient, we also may say, the higher the temperature the greater the average speed of the molecules. Now myriads of molecules at and next to any free water surface, for instance, are moving outwards, and many of them with velocities so great as to carry them into the space beyond and away from that restraining force, whatever it is, that tends to keep them within, and a part of, the body of the liquid.

Outside the liquid the space per molecule is greater, in fact greater without limit, except as restricted by enclosing walls, as in a closed vessel, or by the force of gravity. Hence its free path, the distance it travels from collision to collision, also is vastly greater outside than within. All restraining force of cohesion then is gone and the swarm of free particles is a gas, or, in this case, to be more exact, a vapour, at least at ordinary temperatures—a gas which, at its current temperature, can be compressed to a liquid.

Since great velocity of molecules means high temperature, and small velocity low temperature, and since the greater the velocity of a molecule in a liquid the better its chance to get away from cohesion restraint and become a gas molecule, it follows that evaporation, the changing of a liquid to a gas, decreases the average velocity of the liquid molecules—the faster getting away while the slower remain behind—and hence lowers the temperature of the remaining liquid, or at least keeps it from getting as hot, or hotter as fast, as it otherwise would.

Not only do the molecules in liquid water get away in great numbers and become gas molecules, but also many such gas molecules, in the course of their randomly directed flights, plunge against every water surface with such velocity that they merge with and become a part of the liquid. This is condensation, the reverse of evaporation, more fully explained later. Both processes occur simultaneously, while the circumstances of temperature and pressure determine which shall be the faster of the two. By "evaporation" we generally mean net evaporation, or excess of evaporation over condensation; while by "condensation" we commonly mean net condensation—excess of condensation over evaporation.

Perhaps a duly inquisitive reader may note that nothing at all definite has yet been said about the nature of the forces that keep a liquid just what it is—a liquid—and prevent it from exploding violently into a gas.

To be sure a liquid can be made to explode by heating it to a high temperature under pressure and then suddenly removing that pressure. But under ordinary circumstances a group of liquid molecules does not explode into a group of gas molecules of the same kind, and the problem as a whole still challenges the guidance of clever theories and the test of ingenious experiments.

Although, and confessedly, we soon run out of satisfactory answers to reasonable questions about evaporation we know that it does occur. We know, too, referring now to water, in which we are here particularly interested, that the higher the temperature of the water the faster it evaporates; that the lower the vapour pressure just over the water surface, and therefore also the greater the wind, since it in turn, by remov- ing the accumulated vapour, lowers that pressure, the faster the evaporation; that the greater the extent of the surface the greater likewise, but not in the same proportion, the total amount of water evaporated in a given time; that the saltier the water the slower it evaporates; and that this and that other condition also has its own effects, but generally negligibly small in comparison to those just specifically mentioned.

Although the evaporation of water has been the subject of many studies, as might be inferred from the above list of facts about it the subject indeed of hundreds of investigations no simple and inclusive relation has been found between the evaporation from an open body of water and the different factors that control it. Several empirical equations have been devised to express these relations, but they are far from being satisfactory.

Indeed it seems that the true relation probably is much too complicated to admit of expression in the terms of a simple equation. The best we can do, perhaps, especially in connection with climatological studies, is to obtain evaporation measurements from water pans of the same size and like exposure at favourable localities. Such data do not tell the whole evaporation story, but they have the great merit of being both easily obtained and intercomparable, and they do give a fair idea of the relative evaporativities of the places at which they were obtained.

Ice in any form, from snowflake to glacier, evaporates too sublimes if one insists on using the proper technical term for changing from a solid to a gas without going through the intermediate stage of being a liquid. But it isn't easy to form a satisfactory conception of just how ice, or any other solid, does evaporate. It is agreeable enough to think of the molecules of a gas as darting in every direction among themselves like a swarm of bees on the wing, only more so and altogether blindly; and we can, without shock, liken the molecules in a drop of water to an excited school of minnows, but the molecules of a solid have no such freedom. They are as fixed in position as barnacles on a rock.

Even the atoms, that together make the molecules, keep to their respective places, as we know from X-ray photographs of crystals. Yet they have some sort of motion, oscillatory, presumably, like a pendulum, or the prongs of a tuning fork, for a crystal has temperature just as well as a liquid or a gas, and temperature is a manifestation of the velocities of molecules with reference to each other.

By virtue of this vibration, surface molecules are shaken off more or less rapidly, as determined by the temperature and the kind of substance involved. Many crystals, such as gems of every sort, do not evaporate at all at ordinary temperatures, while others, such as the snow crystal and the crystals of carbon dioxide evaporate freely. The sources, then, of the water vapour in the atmosphere are sublimation from every snow field and ice sheet; evaporation from water surfaces, soil and vegetation; and volcanic steaming from within the earth negligible in comparison to the least of the others.

HUMIDITY

Humidity is just another word for moisture or dampness; but "the humidity" means the degree of wetness of the air, expressed either as "absolute humidity," that is, the quantity of water vapour present per unit volume, or "relative humidity," the ratio of the amount of vapour actually present in any given volume to the amount necessary to saturate that volume at the same temperature. The mass or quantity of vapour necessary to saturate a given volume increases rapidly with increase of temperature, but is not materially affected by the presence or absence of other gases.

Relation of Humidity to Height

Although water vapour gets into, or becomes a part of, the atmosphere at the surface of the earth, it does not stay at that level, but spreads out by diffusion, is mixed with the other gases of the air by convection and turbulence, and drifted about by every wind from the lightest to the strongest. Of course, if the atmosphere consisted of water vapour and nothing else the density of that vapour would be greatest at the surface of the earth (except wherever, as rarely might happen, the temperature decreased very rapidly with increase of height) because the pressure on it would be greater there than at any higher level by the weight of the vapour up to that level.

But as water vapour is only a decidedly minor member in the atmospheric gas association it has but little control of its own fortune, and so is made to behave in ways that often are not its own. One of the most important of these is its distribution with height. It is not always, nor nearly always, densest at the surface, nor does its density decrease with height according to that beautiful law and order that normally applies to the atmosphere as a whole and that would apply, as stated, to the water vapour if it existed entirely alone. In the early morning, for example, after a warm calm night, especially when the ground is wet, or where considerable bodies of water are close by, the absolute humidity, and often the relative, too, is greatest close to the surface, and decreases with height for 50 to 100 feet or more.

After sunrise, however, vertical convection begins, especially when the sky is clear, and frequently an appreciable wind also sets in, and each of these tends to so redistribute the vapour as to make it the same fraction or percentage of the total air at every level from top to bottom of the agitated or mixed up layer, perhaps (and often) 1,000 feet deep. This evidently decreases both the absolute and the relative humidity near the surface and increases each of them in the upper portion, owing to decrease of temperature, to and beyond the saturation point, and thus causes the formation of cloud. When cloud is produced there obviously is left just that much less moisture that can be carried to still greater levels, and so it happens that as the temperature decreases with increase of altitude so too does the amount of water vapour necessary to give saturation. However, the actual amount of vapour over a given place at a particular time seldom if ever decreases nearly so regularly with gain of height.

A column of saturated air, for instance, may rise to a considerable altitude and there spread out, forming a cloud layer. This layer in turn may evaporate and the humid but now clear stratum drift on with air of less humidity in either sense, relative or absolute, both below and above it.

Furthermore, since layers of air of various origins are thus formed at different levels it follows that, in respect to humidity, the atmosphere at any particular time is quite certain to be more or less stratified, no matter how smooth the curve may be that expresses the average relation of the quantity of water vapour to height.

As just implied, we should expect the curve of the average vertical distribution of absolute humidity to be smooth, that is, to show no tendency towards a maximum or minimum at any height, or, in other words, to show that water vapour does not tend either to accumulate at, or to avoid, any particular level. But this natural assumption is in error. There is a tendency for vapour to accumulate at about the limit of daily convection, the level at which rising air spreads out horizontally. On the average this is about a mile and three quarters high during summer in middle latitudes, and much less than one mile in winter, say six to seven tenths of a mile.

Saturation often occurs also at and near the top of the troposphere, as we know from the frequent occurrence there of the cirrus cloud. The amount of the water vapour per unit volume at the very low temperatures, -60° to -70°F, that prevail at this height, 6 to 7 miles above the surface, is very small, even when the air, or space, is fully saturated. That is why the high cirrus clouds are so thin, gauzy and diaphanous in structure and appearance.

As soon as the limit of the troposphere is passed there are no more clouds above to diminish the intensity of the sunshine by day or dim the brilliancy of the stars at night—no clouds here to smoor the sun or blink the stars—except very rarely when either the so-called nacreous, or the so-called noctilucent cloud appears, clouds of wholly unknown origin. The habitual absence of clouds from this upper region is, at first, rather surprising. As stated above, we know from the cirrus clouds that saturation often occurs at the base of the stratophere, and so we should expect the vapour molecules here to wander on up into the stratosphere—to diffuse, as we say, just as odours diffuse from place to place—and thus to bring about saturation at these upper levels also.

We then should except clouds to form at these great heights whenever, for any reason, a distinct drop in temperature should occur. Now, drops (and of course rises also) in the temperature of this region do occur over any given place, drops sometimes of 15°F to 20°F and yet the expected clouds do not appear. Why not? Because, in the first place, there is, at best, very little vertical convection in the stratosphere to carry up bodily the vapour from below, while diffusion, the only other way by which it can ascend, is a slow process. Second, before diffusion can carry up much vapour, or any of it to a considerable distance, the air of this region is sharply cooled and the excess vapour, if any, precipitated in the form of ice needles or spicules that fall down gravity far more rapidly than the vapour from which they were condensed originally rose.

Now, although this cooling at the base of the stratosphere occurs irregularly, it generally does so at intervals of less than a week. Hence distinct clouds do not form, even occasionally, in the stratosphere, if we ignore the rare nacreous and noctilucent clouds, because it is kept exceedingly dry by the frequent formation of an invisible haze at its base. Contradictorily, then, clouds do not form in the stratosphere because clouds do form in the stratosphere—never well up and perceptible, because continually and at frequent intervals, hence low down and imperceptible.

Daily run of Humidity

There are several things that affect the humidity of the atmosphere at a particular place. Some, if not all, of the more important of these are, temperature, convection, nature of the immediate surroundings, wind velocity and wind direction. In general, increase of temperature carries with it an increase in the absolute humidity, through increase of evaporation; and a decrease in the relative humidity, owning to the consequent increase in the vapour capacity per unit volume, a capacity that increases rapidly with increase of temperature. At any rate this is the initial effect. After a sufficient time for much more water vapour to be provided by evaporation the relative humidity may come up to, or even exceed, its former value.

The effect of convection on the distribution of humidity may seem obvious since, as applied to the atmosphere, it com- monly is lower air being pushed up and staying up, with a slight but more widely spread descent of the upper air; or a local descent of the upper air and slight ascent of the neighbouring lower air; or, ideally, a mere interchange of position between the upper and the lower air, the one coming down and staying down while the other goes up and stays up. If, then, in this latter, or ideal, case the lower air at first is more humid than the upper, presently, after convection sets in, the upper certainly will be more humid, one might suppose, than the lower, because the lower air (the more humid) has become the upper, and the upper air (the less humid) the lower. This delightfully simple relation seems, as stated, perfectly obvious, but it is invariably true only for a definition of humidity different from either of those already used.

For these it may or may not be true. If the relative humidity is greater in the lower air than in the upper, then after convectional interchange it will be even more pronouncedly greater in the then upper air owing partly to the increase of its relative humidity, due to cooling by ascent, and partly to the decrease of the relative humidity of the falling air, owing to its gain in temperature. Clearly, though, the lower air may be so dry, with reference to an upper layer, that it has the smaller relative humidity not only before, but also after, they have interchanged places.

Evidently, then, the relative humidity of the lower air may be either greater or less than that of the upper both before and after the convectional turn-over; it also may be less before and greater after, but as explained above, never greater before and less after. Neither does the simple relation, greater before interchange greater after, less before less after, neccessarily hold for absolute humidity (the mass of water vapour per unit volume), since it decreases with ascent and consequent increase of the volume occupied by a given mass of the vapour and similarly increases with descent.

We thus are brought to the alternative of either giving up the simple and seemingly self evident relation mentioned above or of finding some sort of humidity (so defining it) that is not affected by change of temperature or change of volume. Such a definition may, at first, seem impossible, but so long as water is not removed by precipitation these conditions are fully met by "specific humidity," the mass of water vapour per unit mass of the humid air. This particular humidity (as per definition) is not so extensively used as either of the others, relative and absolute, but it is in use, and occasionally, as in the case under discussion, is, by far, the most convenient.

So much, then, for the effect of convection on the humidity of the air— not exactly what one might have supposed, but for that reason all the more interesting. And there are other things that affect humidity. It was stated, a page or two back, that it varies also with the nature of the immediate surroundings. The extreme cases of a small island in midocean and a sand dune in a desert will make this effect of environment practically self evident. On the desert there usually is very little water available for evaporation, so that the air once dry, for whatever reason, normally remains dry for a comparatively long time. In the case of a small island, however, the situation is radically different. Here there is water in every direction that always keeps the lower air, at least, quite humid.

It should be noted in passing, however, that while high humidity is essential to precipitation, that result does not necessarily follow. Catalina Island, for example, is as arid as any portion of the adjacent mainland. High humidity is a factor essential to precipitation but it is not of itself alone entirely adequate. There must coexist also an active process, such as vertical convection, that tends to push the humidity to supersaturation, and would so push it if condensation did not occur. Another of the modifiers of humidity is wind velocity, and it acts in different ways.

It expedites evaporation—the wind, as we know from abundant experience, rapidly dries up wet roads, dries them by increasing the rate of evaporation of the water on them. This increase of evaporation gets moisture into the air at a correspondingly increased rate and thereby affects the humidity. Furthermore, winds are accompanied by turbulence, a mixing of the surface air with all the air for some distance above. In this way the humid air over damp soil and bodies of water is mixed with the commonly drier air overhead, and thus the humidity at all levels in the agitated layer more or less altered.

Also wind direction is an important factor in determining the humidity of nearly all places except, perhaps, those of middeserts and tropical oceans, where its effect is at a minimum and even uncertain as to sign. Evidently the air is more humid when it has just come from over a warm body of water than it generally is when from a dry land area. The sea breeze therefore is more humid, as a rule, than the land breeze, and the summer monsoon winds richer in water vapour than the counter winds of winter.

From the foregoing controls of humidity it is evident that at low levels over the ocean it is quite likely to increase during the day, especially when the wind is light, owing to the addition there by evaporation of abundant fresh vapour, and the slight extent to which it is carried off by convection and mingled with the upper air. Obviously, though, this can go on indefinitely, as on the average it does, only because the accumulating vapour is more or less regularly removed from the place in question. This necessary removal is effected locally and temporarily by the horizontal drift of winds and vertical mixing by turbulence, and eventually completed somewhere else by condensation and precipitation. Over the ocean, too, as there is very little change of temperature between day and night, and but feeble convection, there are no marked humidity maxima and minima at approximately uniform hours, as there are over most land areas. On land, as just implied, the humidity, both relative and absolute, commonly undergoes distinct diurnal fluctuations in the warmer season, not in the colder, with maximum relative humidity of the lower air at the time of minimum temperature and minimum at the time of maximum temperature; and maximum absolute humidity about 8 to 9 o'clock a.m. and again at 9 to 10 o'clock p.m. with, of course more or less pronounced minima at some intervening time, both day and night, as determined largely by wind and convection. During winter the diurnal changes of humidity are comparatively slight since evaporation then is slow, the water being cold, and diffusion incident to thermal convection practically absent. Since change of temperature so greatly affects humidity, and since temperature depends largely on sunshine, it follows that the daily course of humidity is greatly affected by the amount of cloudiness—the greater the cloudiness the less the sunshine, and the less the sunshine the lower the temperature and the less the evaporation.

The Yearly Run of Humidity

Since water vapour is gotten into the air by evaporation, a process that varies directly with the temperature of the evaporating water, and distributed or diffused by horizontal winds and vertical convection, it follows that the annual variation of humidity is very much like unto its diurnal variation. Hence, throughout the first mile or so above the surface of the earth, the absolute humidity of the air is greatest, as a rule, and its relative humidity least, during the summer; and the relative humidity greatest, and absolute least, in winter.

From this in turn it follows that clouds commonly are lower in winter than in summer, both over land and over sea. Of course, too, the annual run of humidity, like the daily, also depends on locality.

Over tropical seas there is but little variation, while in certain other localities the change is all the way from frequent saturation at one season to distinct aridity at another.

Horizontal, or World, Distribution of Humidity

From the fact that all sorts of humidity, the familiar absolute, the more familiar relative and the specific, vary so greatly with temperature, nature of the surroundings and the direction and speed of the wind, it follows that each kind varies also with latitude and longitude as well as with the time of day and from season to season. However, it is not necessary here to consider specific humidity, the weight of water vapour per unit weight of humid air, since it generally varies in the same manner as the absolute humidity, or weight of water vapour per unit volume. The variation of this latter, or absolute humidity, with location is amazingly great. On the most stiflingly hot and humid days in a tropical jungle, for instance, the humidity may amount to even 5 per cent of the air by volume, or 3 per cent, roughly, by weight, while in the coldest places of the world when the temperature is lowest it may be even less by weight than one part in 300,000.

Thus the greatest amount of water vapour that sometimes is in the air where the temperature is quite high is at least 10,000 times greater than that which is or can be in it at temperatures occasionally reached in eastern Siberia. Usually, too, the absolute humidity is quite small over all desert regions, but probably never nearly so small as it necessarily is wherever and whenever extremely low temperatures prevail.

In general, and owing to the very rapid rate at which the amount of water vapour necessary to produce saturation of a given volume increases with temperature, absolute humidity grows rapidly and continuously with decrease of latitude from either pole to the equator, except across the high pressure belts of 25° to 35°, roughly, both north and south—horse latitudes of the oceans, and desert zones of the continents.

Quite commonly the absolute humidity is less along these belts than on either side of them, just as they are also zones of minimum cloudiness and rainfall. The dryness of these zones, in terms of amount of water vapour lacking per unit volume to produce saturation, is more pronounced over the land or desert areas, where there is but little surface and plant water available for evaporation, than along the horse latitudes of the oceans. But strange as it may seem, even these sea areas are arid regions in the sense of having relatively low humidity and scanty precipitation. This is because the air here usually is slowly descending and for that reason commonly rather dry except near the surface.

In addition to these world-wide variations of absolute humidity with location there also are many others of lesser extent. For instance, it decreases in general with distance from the ocean, but not in any regular sort of way. Where the rainbearing winds are nearly all from the same general direction, as they are, for example, over the central United States, they carry, on the average, less and less water vapour, hence exhibit a progressive decrease in absolute humidity, with increase of path over land after leaving the ocean.

This is owing to the fact that the amount of water vapour taken up into the air by reevaporation over any particular area is less than the amount removed from it by rainfall on that same area—less by the contribution of this area to the stream flow, or runoff. But the winds from the different oceans do not all carry equal amounts of vapour, nor do they, as a rule, penetrate equally far into the land, hence the place of lowest average absolute humidity generally is considerably removed from the most inland point. Furthermore, the decrease of absolute humidity with progress of the wind from the ocean often is very irregular.

Thus a coastal range of mountains, where the prevailing winds are onshore, commonly has considerable to heavy precipitation on its windward slopes, while to the leeward the humidity usually be so low as to produce semiaridity for hundreds of miles beyond, unless another mountain range is encountered. And even if such mountains do lie across the winds, and precipitation in quantity does occur on them, the average absolute humidity is not thereby increased, but more or less decreased in proportion to the contribution of the given region to the general stream flow.

The most important things to remember in considering what the distribution of absolute humidity is, and why, are the facts that its possible value, and generally, therefore, its actual value, except in arid regions, rapidly decreases with decrease of temperature; that it is decreased by precipitation; and that it is increased by evaporation. Change of pressure, and change of temperature, by altering the volume occupied by a given mass of air, also change the current value of the absolute humidity.

These same things, mainly, evaporation, precipitation, change of pressure, and change of temperature, also are the factors that control the relative humidity, that is, the current fraction of saturation, or ratio of the amount of water vapour actually present per given volume to the amount necessary to produce saturation at the same temperature of the same volume. It should be especially noted that while the actual temperature is essential to the determination of at least the maximum possible value of the absolute humidity, it is of no importance in fixing the value of the relative humidity.

Since the vapour capacity of a given space or volume—the maximum amount of water vapour that can exist in that space in the presence, to be exact, of a flat surface of pure water-depends on temperature only and rapidly increases with increase of temperature, it is evident that, with a given amount of water vapour, short of that necessary to produce saturation, present per unit volume, any decrease of temperature will increase the relative humidity.

This is owing partly to decrease of the volume, the pressure remaining the same, occupied by a given quantity of the vapour and partly to the decrease of the vapour capacity per unit volume, both incident to the decrease of temperature.

Conversely an increase of the temperature, other things remaining equal, decreases the relative humidity, and so too does decrease of pressure, temperature remaining constant, owing to the corresponding increase of volume. From these fundamental principles it is practically obvious how the relative humidity must vary with change of place. Thus, tropical winds on their course to higher latitudes normally become colder, and therefore gain in relative humidity even to saturation and the production of fog and cloud. Similarly the relative humidity of winds from higher latitudes generally decreases, owing to increase of temperature, except where evaporation is particularly rapid. Again, in the summer, relative humidity commonly decreases with progress inland over a continent, owing partly to increase of temperature and partly to an actual decrease of the humidity in the air.

In the winter the relative humidity generally is greater over continents than over oceans, owing to the comparatively low temperature then of the land surface. Naturally, too, the relative humidity is especially low over arid regions, and this condition extends in a measure to the oceans, where the relative humidity averages about 75 per cent along the horse latitudes, and 80% over the rest of the ocean areas.

CONDENSATION

Condensation, or the changing of water from the gaseous state to the liquid state, is exactly the reverse of evaporation. And just as the latter, that is, evaporation, makes the space into which the vapour passes humid—makes the air humid (part water) we commonly say—so obviously its reverse, condensation, always reduces the absolute humidity, or amount of water vapour present per unit volume.

These two processes, evaporation and condensation, are the fundamental controls of humidity, every form of which, so long as the temperature and the pressure are kept constant, is increased by the one and decreased by the other. It, therefore, is desirable to know something of the conditions under which this important phenomenon, condensation, occurs, and to formulate some idea as to how it occurs.

It is now well established by evidence fascinating and convincing, but too long to include here, that each pure gas, hydrogen, oxygen, water vapour or what not, consists of equal discrete particles so infinitely small that if 500,000 people were each picking them out at the rate of 100 per minute, day and night, it would take them 1000 years to accumulate particles enough to make a volume of the air we breathe as large as the head of a pin. And yet we know to closer than one part in a hundred the weight or mass of each of these particles, and how any kind, oxygen for instance, differs from any and every other kind. We know, too, that all these particles are in swift motion, on the average, among themselves with, seemingly, all the irregularities of a swarm of bees, and more besides. Yet in spite of this apparent utter confusion as to direction, and wide range of velocity, certain surprising regularities or laws do obtain. Thus the average velocity of the molecules varies directly as the square root of the absolute temperature, inversely, as the square root of its mass. In a mixture of gases, such as constitute the atmosphere, the average energy of each molecule is the same as that of any other—whatever the lighter ones lack in mass they make up for in speed. And that speed is astonishingly great, in fact at ordinary temperatures it is around 1500 feet per second, or say that of an ordinary rifle bullet, while the velocity of the water molecule in the is approximately 1900 feet per second.

But regardless of their great velocities the molecules of a gas, or anyhow much the greater portion of them, seem to come to rest when they impinge onto a solid or liquid surface. That is, at such times they at least momentarily condense where they strike. If the temperature of the object struck is high they are quickly knocked off again, but if its temperature is low, if, in other words, its molecules vibrate slowly, the chance that the arrested molecules will promptly leave it is correspondingly less. Hence, at a certain temperature which we call the dew-point more moisture collects on a water surface, and also on many other objects, than leaves it.

This is condensation. When this condensation occurs on foreign particles in the atmosphere, however minute they may be, the result is a fog or cloud. When every thing of this dust-like nature is taken out of the air, by filtering it through a tube stuffed with raw cotton, or otherwise, fog does not form immediately the dew-point is passed, nor indeed under any degree of supersaturation that possibly can occur naturally and in the open. We know, therefore, that if the atmosphere were pure air, in the sense of containing nothing that could be mechanically filtered out of it, never a fog nor cloud would form, nor rain nor snow, only evaporation from the surface, and condensation again somewhere on the surface—evaporation chiefly in the warmer regions and condensation in the colder.

The atmosphere, however, is not pure, but filled with countless myriads of condensation nuclei of many kinds—kernels as it were about which water shells can and do form, immediately the state of saturation is passed. Though but an infinitesimal part of the ordinary outdoor air they nevertheless are absolutely essential, indeed weather vitamins without which there could be no rain and no snow.

Clouds Yield Rain

The above may describe the making of a cloud droplet, but a cloud droplet is a long way from being a rain drop. Indeed it takes about a million of the former to make one of the latter, and besides the droplets have but little tendency to spontaneously and rapidly unite, as is evident from the fact that clouds often float along for hours and even days without yielding a drop of rain. On the other hand, rain often does occur and so we wonder why clouds sometimes produce abundant rain and at other times none at all. They look quite alike on the two occasions, as like as peas in a pod, and are alike except that when rain is falling a lot of relatively large drops are present while none are present when no rain is falling.

But that is the point. Why are the clouds not always the same, always consisting entirely of minute droplets or always containing also rain drops? Well, as stated, there is no important difference between these clouds except in respect to the detail of the size of some of the drops, but there is a marked difference in the movement of the air in which they occur, and in this rests the secret we are trying to find. Clouds from which no precipitation, rain or snow, is falling, are floating along in air that has very little vertical movement, whereas there always is appreciable, and at times even great, velocity upward of the air in which rain drops or snowflakes are forming.

Now, water drops fall through still air with a velocity that varies with their size, from about one tenth of an inch per second in the case of the cloud particles, to some 25 feet per second for the largest possible rain drop—one fifth of an inch in diameter, approximately. Clearly, then, whenever the movement of the air has an upward component, clouds will form as soon as expansion has brought the temperature to the dew-point, and will so continue to form in the ever fresh air at the same level as long as that particular stream continues to rise. Yet, despite this constant formation of cloud not a droplet can fall until by growth from condensation, or coalescence, or both together, it has become large enough to overcome by its weight the upward drag of the rising air. And, further, since more and more cloud is formed, and since well up in the cloud the velocity of ascent grows less and less with increase of height, and finally ceases altogether, it follows that after a time there must be a falling of drops of appreciable size; in short, a rain, and at such rate as to equal the fresh supply of cloud, or new condensation.

Other Forms of Precipitation

Rain is not the only form of precipitation: Snow.—When the condensation occurs at temperatures below the freezing point the usual result is a snow crystal, and when, in the course of their fall, many such crystals get entangled together into a single mass the result is a snowflake. Sleet.—Sometimes rain falls into a freezingly cold layer of air only a thousand or so feet thick and reaches the earth as frozen pellets, or sleet of the variety that rattles when it hits a windowpane.

- *Hail:*—On the other hand, in a vigourous thunderstorm rain drops frequently are carried to such great heights that they freeze, and often grow to considerable size before they overcome the lift of the rising air and fall to the ground as hail stones. 188 Encyclopaedia of Environmental Science
- *Graupel:*—Still another form of precipitation is the rather unfamiliar graupel or soft hail—white, granular bits of ice roughly spherical in shape and an eighth of an inch, more or less, in diameter. It may be formed in subcooled clouds, that is,

clouds of water droplets at temperatures below the freezing point, much as rime, or granular snow tufts, are formed on the windward side of exposed objects in a drifting, subcooled fog. Certain particles here and there in the cloud may, and undoubtedly do, fall faster than most others, and grow by capture as they fall. If, now, the cloud is subcooled the droplets are certain to freeze immediately they come into contact with the falling larger mass and adhere to its underside, thus giving it, as it turns every which way in the course of its fall, an approximately spherical form.

- *Fog drips.*—The drippage from trees and other objects when a dense fog drifts through them. Dew and frost are intentionally omitted from the above list of precipitation form, because they do not fall from the clouds as do rain and snow.
- *Qiantity of precipitation.* What is the quantity of all this precipitation? Well, for the whole world there is lots of it. There is enough humidity in the atmosphere all the time to make a layer of water at least an inch deep over the entire earth, while the average rate of the total precipitation, and of course also the rate of evaporation, is approximately 16,000,000 tons per second, enough to cover the whole earth in the course of a year with a layer of water three feet deep.

Atmosphere Usually is not Saturated

Evidently this precipitation furnishes all the water used in plant growth, and all that runs off as stream-flow except, to be exact and if geologists are right, a portion of that negligibly small amount that comes from gysers and certain hot springs—a portion furnished, some hold, by the primordial rocks themselves. But there is another important thing that results from the occurrence of precipitation that is not so obvious. It is the fact that in this way the atmosphere is kept from being everywhere and at all times intolerably humid, indeed saturated, or very nearly so. Every rain and every snowfall takes that much water out of the air—dries it to that considerable extent and thus keeps the humidity generally far below the saturation point, and therefore normally within the bounds, not of endurance only, but of actual comfort.

DISTRIBUTION AND CHANGES OF ATMOSPHERIC PRESSURE

Although atmospheric pressure, or weight of the air per unit horizontal area directly above the place where the pressure is measured, is not a weather element of any particular importance so far as our feelings are concerned, such as temperature and humidity are, nevertheless it is the element that more than any other locates the weather of today and tells us what tomorrow's will be, and therefore is of great importance. Latterly this importance has appreciably decreased, however, because other methods have come into prominence for finding the positions, magnitudes, speeds and directions of travel of the contrasting masses of air whose conflicts and whose passage keep the weather in ceaseless ebb and flow from fair to foul and foul to fair. Still these other methods have not yet been sufficiently developed to justify ignoring the distribution of atmospheric pressure, and how it is increasing here and decreasing yonder, as aids to a foreknowledge of the weather of the morrow. Neither is it at all likely that so great a development and use of these other means, while entirely possible, will be effected at an early date, if indeed they ever are so developed. Therefore a knowledge of the distribution of atmospheric pressure, the near equivalent of the distribution of the air itself, and of its changes, still are, and for a long while must remain, matters both of great practical importance and of keen theoretical interest.

Barometer Corrections

Since the pressure of the air at any particular time and place is, as stated, the weight of all the atmosphere above a horizontal unit area then and there it follows that a knowledge of the values of this pressure over a given region, however extensive, tells us, when the proper corrections have been made, just what the distribution of the air itself is over that same region. But what are these correc- tions? That depends on the kind of instrument used to measure the pressure and on what we want to find, pressure or quantity of air, for we must not forget that the same quantity or mass of air, as also of other things, varies in weight, or pressure, with position not with every change of position but with most changes.

But we can think easier and straighter if we consider one thing at a time, and know exactly what that one thing is. Let us first, then, compare the masses of air over different places. We can do this by means of a mercurial barometer which is only a device for balancing the pressure exerted by a column of mercury at its bottom, tending to make it flow out, against the pressure there of the atmosphere that keeps it back in. Clearly the pressure exerted by the mercury at the bottom of the column, when corrected for the capillary action of the tube, is proportional to the product of its density by the height of the column by the force of gravity, or pull oil a unit mass by the earth, at that place.

Similarly the pressure there of the atmosphere is proportional to the weight, or mass times the force of gravity, of all the air, if still, above that level (the level of the bottom of the mercury column) per horizontal square inch, or other specified area. When the air is in motion, however, as it always is, a further correction of the reading of the barometer is indicated, since the weight (downward push) of a given mass varies with the direction and speed of its motion. But since the value of this correction (never known accurately) at most is negligibly small it will not be discussed here just mentioned for the sake of completeness. Similarly the effect of the variation of gravity with height also is negligible. In this case then we may consider sauce for the goose as sauce for the gander too, in the sense that gravity applies in equal (almost) measure and exactly alike for both the mercury and the air. Hence the mass of the air above any given horizontal area, such as a square foot, say, at the level of the mercury in the barometer basin, is proportional to the height of the mercury column, or reading, as we say, of the barometer, corrected for whatever slight difference in that height there may be due to the pull, or capillary action, between the mercury and the wall (glass) of the barometer tube, and corrected also for the density of the mercury corrected to a common or standard density, commonly that which it has at the melting point of ice, 32° Fahrenheit, or 0° Centigrade.

In this way the readings of all the mercurial barometers are made strictly intercomparable, and the ratios of their several readings to each other exactly the same (to within an unknown but always negligible amount) as that of the masses of air above them, so that a geographic map on which these values, if simultaneous, are written, or otherwise represented, is also a map of the distribution of the atmosphere over that same region at the time the readings were made. To obtain the instantaneous distribution of the air the barometer readings necessarily must all be taken at that particular time, whatever it is, 8 A.M. 75th meridian time, or what not.

If, however, aneroid barometers are used instead of mercurial for obtaining the mass distribution of the atmosphere, then their readings, which indicate pressures directly, must be corrected for the local values of gravity, because the atmospheric pressure, being the weight per unit horizontal area of a vertical column of air, varies directly with the force of gravity which, in turn, increases with latitude. This quantity, or mass, distribution of the atmosphere is of considerable importance, especially through its influence on the rate of transmission, in or out, of radiation. A thin atmosphere and a thick one are to the earth much like a sheet and a blanket, respectively, to one sleeping in the open.

Although this mass distribution of the atmosphere is very important as a factor that helps to make the weather, especially as to temperature, what it is, still very little use ever is made of it. The distribution of pressure, or push, in the air is on the contrary frequently, carefully and extensively plotted because it is the unequal distribution of push on the atmosphere (or pressure, for they are equal to each other) that makes the winds to blow and thereby, in large measure, but not wholly, change the weather of one place by bringing to it the air and the weather of another. Clearly, then, a knowledge of this distribution must be, and is, a great help to the fore- caster when judging of the weather of the morrow and the day after. That is why the barometer is so extensively used in making meteorological observations and used almost exclusively for measuring the atmospheric pressure.

Now the pressure exerted by a column of mercury, the column in this case that is balanced by the pressure of the air, is proportional to the height of the column (corrected for capillary drag), to the density of the mercury (which in turn varies with the temperature) and to the local force of gravity. Hence, to compare the pressures at different places, one with another, it is necessary to correct the readings of the barometers to what they would be if all the instruments had the same temperature, say the temperature of melting ice, and were actuated by the same force of gravity its value at sea level at latitude 45°N or some other agreed-upon value.

The readings of aneroid barometers require no corrections for this purpose since, as previously stated, these instruments measure pressure directly. To sum up: If we are measuring the mass distribution of the air the readings of the mercurial barometer must not be altered for gravity differences, while those of the aneroid must be so changed.

On the other hand, when it is the distribution of pressure we are finding, the element we nearly always seek, it is the readings of the mercurial barometer that must be corrected to what they would be under some common force of gravity and not those of the aneroid barometer—they are true pressure values, if the instruments are correct, as they stand, and therefore must not be changed.

The next and final step is to reduce all these pressure values, or corrected readings of the barometer, to what they would be if taken at some common level. This is done because the travel or flow of the air, due to difference of pressure, is, of course, along the route of least resistance, which route, as a rule, is over a nearly level surface, that is, a surface everywhere normal to the direction of the force of gravity, such as the mean surface of the ocean. Actually the direction of least resistance to a sample of air through the surrounding atmosphere, the direction indeed of no resistance at all, save alone that incident to viscosity, is along what the physicist calls an isentropic surface. However, this exact and refined treatment of the subject will not here be followed to its conclusion, though to do so affords both surprises and mental satisfaction.

But let us have as little boggy ground as possible under our mental feet. Why is the level surface usually the one of least resistance to a sample of moving air? Because ordinarily the change of temperature of the atmosphere with change of height is such that work would have to be done on the sample to carry it to a greater height or push it to a lower level, but no work (save that due to viscosity which is the same in all directions) to move it horizontally.

Normally this distribution of temperature is such that a sample of air taken upward would cool, incident to its expansion under a decreased pressure, to a lower temperature than that of the then adjacent air. Being at the same pressure as this adjacent air but cooler, it also would be denser, so that an actual lift would be required to carry it above its initial level. On the other hand, if taken downwards the compression due to the increased mass of air above it would warm it to a higher temperature than that of the other air at its new position.

But being now warmer than the adjacent air, and having the same pressure, it also must be lighter. Hence, normally, only an actual downward push can force a sample of air at any height to a lower plane. In short, air normally can be taken out of its own level only at the cost of work, work against gravity if pushed up, and work against the buoyance of the denser air beneath if forced down.

Clearly then no work against gravity, nor any against buoyancy, is required when, and only when, air is moved over a level surface, or, to be exact, over an isentropic surface—a surface which, as already explained, usually is very nearly level. Obviously any direction over this surface is a direction of least resistance.

Therefore if we would know the magnitude of the cause of the winds, and from that magnitude infer their strength, we must find the values of the atmospheric pressure at the same level at various places over the region in question.

If we had a great number of these pressure readings somewhat evenly distributed over an area of considerable extent, say the size of the United States, or larger, we then could draw on a map of that region a line connecting all the places that, at a particular instant, had the same pressure. This line, according to present practice, we would call all "isobar," or line of equal pressure. By drawing a number of such isobars, corresponding to small but constant differences in pressure, we construct a map or diagram from which the horizontal pressure gradient, or horizontal push per mile, say, can be found. Note that no two isobars (lines of equal pressure) can cross each other. Such crossing would signify two different pressures at the same time and place, a manifest impossibility.

The direction of the flow of the air under these unequal pressures alone would be horizontal, as previously explained, and along the course of the shortest distance between the lines of equal pressure, that is, normal to these lines. A sample of air would not flow off slopingly to this normal to the isobars, though that too would be going from a greater to a lesser pressure, because every component of pressure directed to the right along or parallel to an isobar is exactly balanced by a component of pressure in the opposite direction. The only unbalance is strictly normal to the direction of the isobars, and therefore this is the direction of flow, or would be, if there were nothing to disturb or prevent such flow. As a matter of fact this flow is modified by the rotation of the earth, and so profoundly modified as to change it from the direction of the push to the right angles thereto, except in the lower air where the drag of the surface makes the change in direction less than a right angle. But this fascinating paradox must be passed by. Its explanation would be out of place here; besides only the language of mathematics properly fits it anyway.

Since the movement of the air under differences of pressure is horizontal, or very nearly so, as just explained, it is clear that our isobars must be drawn for a common level if we would know from them what the values of the pressure gradients are that are making the winds of the moment to blow, fixing the intensity of a passing storm and largely determining whither and how fast it shall go. But what should be this common level, and how can we find the value of the pressure there?

These questions are old; their answers would be new. For the mariner they are simple enough: sea level is the place, and read your barometer where you are (on that level) is the way. But for the landlubber it is a different and awkward matter, since in general no two of his observing stations are at the same level. For the purpose of weather forecasting one can chart on his map the departures of the barometer readings of the several stations from their respective normal values for the season, or day, in question, and then connect by a smoothly curved line the places of equal, like departures.

Such a map can be, and has been, used with fair success, but it never became a general favourite, though it is the only pressure map that deals with known accurate magnitudes. One thing that renders such a map less helpful to the forecaster than at first one might expect it to be is the fact that the normal, or average, pressure for the given month, say, is one thing for weather of one kind and another for weather of a different kind. In short, the general average or normal for all days is not applicable alike to all sorts of weather conditions— the normal itself, or the observed departure from it, needs to be corrected for each special reading, but by how much we do not know. The pressuredeparture map therefore is not entirely satisfactory and can not be made so.

Another promising procedure, and one that has been seriously attempted, is to use the pressures that exist at some definite height above sea level, half a mile, one mile, or what not. Wherever this level is above the surface, surely at that level there is air. Surely, too, instruments can be carried to that level and the actual pressure there measured with considerable accuracy. However, this direct manner of obtaining the desired pressure values is now too expensive to be practicable. Therefore the method used has been to compute the pressure at a given height from the pressure and temperature at the surface with a general formula and then correct the value found according to the circumstances, especially as to season and direction of the wind, as empirically deduced from data obtained under various conditions by kite, balloon and airplane. But at best more or less doubt as to accuracy always attaches to the values thus obtained. Hence weather maps that ostensibly show the distribution of atmospheric pressure over a horizontal surface at some definite height above sea level have not yet come into general use.

There remains, in this connection, one other obviously possible procedure, the one that has been, and now is, all but universally and solely used. This is, to reduce each pressure reading made on land to what it presumably would be at sea level immediately beneath the point of observation if all below that point, and round about, were not rock and soil, but air in that state and condition appropriate to the location, season and kind and distribution of the weather. The calculation, only approximate of course, is based on the known height of the station above sea level, and the observed temperature and pressure of the air, and then empirically more or less corrected. The chief source of error is the local surface temperature. Sometimes this is abnormally high and sometimes abnormally low. In the first case the reduction to sea level is made on the assumption that the hypothetical air beneath the place of observation is correspondingly warm and light, and in the second that it is correspondingly cold and heavy.

The one yields a pressure too small and the other a pressure too great. Occasionally this error is so big that even the sense of the pressure distribution is reversed—actual cyclones and anticyclones, as indicated by the direction of the winds, changing by reduction to anticyclones and cyclones, respectively, on the map. Pity then the forecaster. But don't pity him overly much. He doesn't deserve it. He just winks the other eye, having in mind any poor devil trying to learn the game from such charts, mentally alters the map in accordance with his background of experience and all data available, some of which are never charted, and goes ahead serenely, if not always confidently. At his own game, and with loaded dice, of course he beats the innocent novice—until he, too, gets wise.

The one thing we definitely know about all reductions of the barometer to sea level, or any other level, is this: They invariably are in error. Generally, however, they are usable, and we do not know of any practical way to improve them. We know too that a "level" surface is not the correct surface on which to draw isobars; but then the surface that is the proper one, the isentropic surface, does not "stay put," as we say, but perpetually so warps and waves as to make pressure reduction to it quite out of the question.

As a rule, however, this correct surface is, as previously stated, nearly level. Hence, and because reduction of pressure to a common height usually is only slightly erroneous, it follows that the sea-level map meets the needs of the weather forecaster fairly well, and better indeed than any other that as yet it has been practical to construct.

In this connection let us confess a little inconsistency and confusion of practice in our sea-level reductions. We reduce the pressure, or barometer, readings to sea level, but we leave the temperatures unchanged. Hence our weather maps show sea-level pressures and "mile-high" temperatures (themselves often exaggerated by surface conditions) side by side. This for the forecaster. When the climatologist sums the weather data all up he commonly reduces both temperature and pressure to sea level. Consistency, thou mayest be a jewel, but art not always prized.

GENERAL DISTRIBUTION OF PRESSURE

Since the atmosphere is so exceedingly fluid, moving horizontally under the slightest push, one might suppose that its sea-level pressure would everywhere be the same and therefore its mass distribution substantially equal the world over. However, no such uniformity obtains. In fact, this pressure varies from place to place at any time and from time to time at any place, seasonally, diurnally, semidiurnally, and irregularly in different manners. And there are two fundamental causes of all the larger and more important of these changes of pressure; namely, differences in temperature and the rotation of the earth.

Because, on the average, the atmosphere for miles above the surface of the earth (not all the way out, however) is warmer in equatorial regions than at middle to high latitudes, it there is most expanded and from there has most overflown towards either pole. Hence for this reason alone, if there were no other, the sea-level pressure of the air would be continuously lower over the tropical portion of the earth than over the colder parts on either side.

But there also is another important factor contributory to the final result, namely, the rotation of the earth, in consequence of which the circulation, just mentioned, of the air between the equator and the poles seldom moves with the same speed and in the same direction as the surface of the earth itself, though this is the state, a dead calm, towards which surface friction and turbulence always and everywhere tend to bring the air. The normal resultant of this tendency, rarely anywhere satisfied, is a system of more or less continuous winds from easterly points between the latitudes 30°N and 30°S, roughly, and a system of westerly winds in each hemisphere between the latitudes 30° and 60°, or thereabouts. The winds beyond latitude 60°, especially those of the southern hemisphere, again are more or less from the east.

But what has this to do with the pressure of the atmosphere? Much. The bulge of the earth is just such that a frictionless object on its sea-level surface if once at rest would remain at rest in perfect equilibrium, and such would have been the case if the rotation of the earth had been either greater or less than it now is. In the first case the bulge would have been correspondingly more pronounced and in the second less pronounced. If therefore our frictionless object were given a velocity eastward it would find the earth insufficiently bulged, so that to it equatorwards would be down hill. Similarly, if it were given a velocity westward its down hill then would be in the direction of the adjacent pole. Hence the winds of the tropical regions, being from the east, tend to drift off polewards, those south of the equator towards the south pole and those north of it towards the north pole. Therefore in this mechanical way, as well as by virtue of the expansion and consequent overflow of the relatively warm air, a belt of low sea-level pressure is established and maintained along, and for some distance on either side of, the equator.

Also, the poleward push of the tropical winds is roughly balanced by the equatorward urge of the mid-latitude winds- winds prevailingly from the west—from which there results two belts of high pressure, one in the northern hemisphere with its axis roughly along latitude 30°N and the other similarly situated in the southern hemisphere. Beyond each of these belts of high pressure in turn, and for exactly similar reasons, there is a belt of low pressure with its axis at about latitude 60°.

The one in the southern hemisphere is by far the better developed of the two. All along it the average pressure is decidedly lower than it is anywhere else on the globe. The current pressure there, however, undergoes frequent and great changes owing to the passage of storms as presently to be explained. Beyond these circumpolar belts the sea-level pressure again increases, as is evident from the fact that there is a more or less continuous outflow of surface air from each of the polar regions.

In respect, then, to the major features of the distribution of the atmosphere over the earth there are seven distinct great regions: three where it exists in smaller amount than the average for the world as a whole, namely, the equatorial belt and the two 60° latitude circumpolar belts; and four, the two high pressure belts of latitudes 30°N, 30°S and the two polar caps, where it occurs in amounts larger than the average.

Centers of Action

Not only are there seven permanent divisions, as just listed, in the distribution of the atmosphere and its sea-level pressure over the earth, but three of these divisions in turn have important semipermanent subdivisions, which, owing to their connection with the direction and strength of winds, and course, intensity and frequency of cyclonic storms, commonly are called centers of action.

Several of these are the peaks of high pressure that normally exist near the eastern ends of the ocean portions or segments of the high pressure belts. These belts are far better developed, and undergo fewer and smaller changes over the oceans than on land, but even here they are not of equal intensity from end to end. Near the eastern limit of each of these links the pressure normally is much greater than it is elsewhere, a circumstance quantitatively related to the strength of the surrounding winds. As this pressure maximum shifts in position and intensity so do the winds of that general region also vary in place and strength. Evidently there must be, and there are, five of these highpressure centers of action, one each in the North Pacific, South Pacific, North Atlantic, South Atlantic and Indian Oceans, respectively.

Besides these there are two recognized low-pressure centers of action, namely, the Icelandic low, southwest of Iceland at about latitude 60°N and the Aleutian low, just south of the eastern Aleutian islands and near latitude 55°N. The former of these, the Icelandic low, though shifting in position, and often being replaced for short intervals by high pressure, or an anticyclone, nevertheless statistically, or on the average, is present month after month and year after year as determined by the distribution of temperature over and round about that region.

The Aleutian low, on the other hand, is only a winter phenomenon when the land areas round about are much colder than the water where the prevailing low pressure occurs. In the summer time when the land areas are comparatively warm, low pressure is not a normal feature of the ocean in the general region of the Aleutian Islands or the Gulf of Alaska. Both these "lows" also frequently are called centers of action since many migratory cyclones either start from them, or, at least, come by way of them from regions beyond. There are no definitely known counterparts to these cyclonic, or low-pressure, centers of action in the southern hemisphere. Perhaps, however, they do exist, one over the Ross Sea where the surface temperature obviously is higher than that of the mostly surrounding great glaciers and snow fields, and possibly another, but much feebler one, over the Weddell Sea.

Pressure Changes

As every one knows, where the air is being heated it expands and where it is being cooled it contracts. Therefore the atmosphere overflows from the summer hemisphere onto the one where winter prevails, and thereby causes a difference between their respective quantities of air and values of their sealevel pressures. The change is small from winter to summer, only about one part in three hundred of the whole. However, in quantity of air this comes to, roughly, 10,000,000,000,000 tons which are shifted across the equator every six months from the warming to the cooling half of the world.

Not only is there a world-wide shift of the atmosphere with the seasons, but also there are like-caused, similar shifts of the air with the seasons between the oceans and the continents of each hemisphere. In winter the average temperature on the land is much lower than that of the adjacent oceans, and in summer much higher. Hence their air pressures also vary from season to season, being greatest over the continents during the colder season and least during the warmest, and over the oceans greatest in summer and least in winter. This difference between land and sea pressures varies widely from place to place, but on the average is greater than that between the summer and winter pressure of either hemisphere (northern or southern) as a whole. The equatorial belt of low pressure varies a little in position owing to the annual shift of the axis of maximum temperature, or thermal equator—nearly always and everywhere a little north of the geographic equator. Evidently the shifting of this belt necessarily entails some change, though small, in the air pressure of the region covered by it. The high pressure belts also shift north and south, and through several degrees, thereby causing other annual swings in the pressures of many places. The Icelandic low, as previously mentioned, is much stronger in winter than in summer, while the Aleutian low exists only through the winter. Whether there is much change in the low-pressure belt around the border of Antarctica is not entirely certain, owing to lack of adequate observations through the year in that part of the world. Presumably, however, there are some annual changes in the depth and position of this belt also as there are in the intensities and locations of all the others.

In addition to these annual swings in position and intensity of the belts of high and low pressure, there also are marked occasional variations in the locations and strengths of their centers of action, the peaks on the ridges of high pressure, and the depths in the circumpolar valleys of low pressure. When marked shifts of these centers will occur can not yet be accurately foretold, nor indeed do we know in detail any specific chain of circumstances that produces such a result. We do know, though, that they often cause, so long as they last, a great difference in the weather of neighbouring lands, and at times even of places far away.

For this reason the centers of action, both the cyclonic and the anticyclonic, deserve and have received much study. We do not yet fully understand them, at least not so well that we can know in advance when and where they will occur nor how decided they will be, but that is no reason for giving over our study of them; rather is it a challenge to make our study all the more persistent and thorough. We should here emulate the Frenchman who, with the courage of his convictions, said that though we do not understand woman, and may never understand her, we are not for that reason going to give her up. An even greater puzzle among the many pressure changes of the atmosphere are the occasional unheralded surges towards greater pressure or lesser pressure of roughly equal amount simultaneously, or nearly so, over an area of continental size. It may last for days and then disappear as mysteriously as it came. So far as the weather is concerned these surges appear to be of little importance, but until we do understand their cause and know their effects they will remain a taunt of ignorance that leaves us mentally ill at ease. Possibly, too, when we know them better we shall find them more important than we ever before dreamed they could be.

The most important of the pressure changes, due, we must remember, to changes in the amount of air overhead, are those that in middle latitudes occur on the average at least once a week, and often amount to fully three per cent of the whole. These changes are incident to the passage of those systems of wind, and their accompanying phenomena, which we call cyclones in which the pressure is relatively low, and anticyclones in which it is comparatively high.

In most tropical regions, but not all, nor anywhere close to the equator, cyclones, often very intense, occur, but never an anticyclone. The anticyclone is the harbinger of fair weather, and the cyclone, wherever it occurs, the bringer of cloud and commonly also of rain or snow according to the current temperature. The forecaster locates on his map these systems of wind and weather, and from his understanding of the subject and intimate familiarity with such maps, quickly estimates what the weather will be a short time ahead, usually 12 to 36 hours, at each place over his territory, large or small. In some cases he makes a general forecast for at least a week ahead, and in others, as for the aviator, a very detailed forecast for only the next two or three hours.

There still are a few more irregularly-occurring changes in the pressure of the atmosphere that should be mentioned. The precipitous drop in the pressure, amounting in some cases to probably one tenth of the whole, that occurs with the central passage of the destructive tornado. This extreme drop in the pressure is owing to the centrifugal force incident to the violent rotation of the air about the axis of the storm. The irregular fluctuations of atmospheric pressure that accompany the earlier stages of a passing thunderstorm, due, in large measure, to irregularities in the downward velocity of the air in the front portions of such storms. The minute fluctuations in pressure of 5 to 10 minutes period that sometimes occur when a shallow (a few hundred feet deep) surface layer of cold air is overrun by a wind of warmer air which throws the former, or cold layer, into waves or billows, much as winds make waves on the ocean. (4) Finally, if we think to have done with all these irregularities in the atmospheric pressure at one full swoop by the use of an automatic registering apparatus of exceeding delicacy, such as has been constructed and tried, we shall find that we have but multiplied our difficulties beyond endurance, for minute changes are continuously occurring that appear to be wholly without order or regularity of any kind.

But even all the above changes, annual and irregular, of the readings of the barometer do not exhaust its known variations. Indeed only a few years after the invention of the barometer in 1643 an article was published in the Philosophical Transactions of the Royal Society of London on the twice daily symmetrical swing of the atmospheric pressure, which in tropical regions amounts, from maximum to minimum value, to roughly one part in 150 of the total pressure.

The amount of this pressure change decreases with increase of latitude, slowly at first and then so rapidly that it becomes uncertain beyond latitude 60°. Its maxima occur at about 10 o'clock, morning and evening, and its minima at 4 o'clock, both A.M. and P.M. There appears also to be a variation of the barometer with an 8-hour period. Furthermore there is a daily variation of the pressure from maximum, in valleys and other low places just before sunrise when the air is coldest and most contracted, to a minimum when the air is most heated and expanded.

On mountain tops, and other places of considerable elevation, the pressure is minimum and maximum when down below it is maximum and minimum, respectively. Still further variations in the surface pressure of the atmosphere are caused by the tidal actions of the sun and the moon. Those due to the action of the sun occur twice daily, but are very small, having a total variation even in tropical regions where it is most pronounced of only about one part in 40,000 of the whole pressure. It therefore can be detected only by averaging a long series of observations hour by hour so as to eliminate irregular variations and then subjecting the result to rigid analysis. As stated, there also are similar, but about two and a half times larger, variations in the pressure incident to the air tides caused by the moon. Even these no novice at the game would ever detect. Lastly—not the last that could be mentioned, but the last that here will be mentioned—there are the annual solar tides due to the swing of the sun north and south of the equator, wholly lost, perhaps, in the much greater thermally induced swings of the atmosphere in the opposite directions; and the lunar-monthly tides caused by the drifting of the moon from one hemisphere to the other across the equator.

As just stated, not all the changes in atmospheric pressure, all the comings and goings of the air but enough of them, no doubt, to make it clear that the ways of the air are devious and strange—full of puzzling problems that it is an intellectual delight to solve, and in every sense profitable to understand.

WIND

"Never mind the weather so the wind don't blow," runs the old song, turgid with the wisdom of experience, that our granddads used to sing in their boyhood days. But the winds do blow—blow comfort when gentle, annoyance when strong, and destruction when violent. Naturally, then, they were of vital importance to our ancestors of the tree tops, and important even to our great-great-great-to-the-nth-generation grandfathers who went to the other extreme and lived in caves, for whatever the shelter they found within the earth, surely the game that hunger compelled them to chase had to be captured on the outside in the realm of Boreas and all his blustery kin.

Nor have the artificial caves we now make of all sorts of materials, from stone and mortar to stubble and straw, and place on the surface of the earth, improved the situation one whit, for they, too, even more than natural caves, leave us subject to stress and storm.

There remains the same compulsion to leave them for the open betimes, nor are they always proof against every wind that blows.

And so it has come about that ages ago man asked: "What is a wind?" "Where does it come from?" "Where does it go to?" "What caused it?" They asked a lot of other questions too, but these will do for the present. To the first of these questions, "What is wind?" a Greek philosopher replied: "Air in motion." No better definition of wind than this ever was made nor ever can be try it. The next questions proved to be more difficult. In fact until yesterday, relatively speaking, they were either given up wholly as unanswerable". The wind bloweth where it listeth, and thou hearest the sound thereof, but canst not tell whence it cometh, and whither it goeth"—or else answered in an entirely fanciful and irrational manner, however pleasing the concept and beautiful the poetry that adorned it.

According to one such concept a demigod, Aeolus, had charge of the winds, which he kept confined in a great cave, and from which this or that one was let out as whim, fancy or need might suggest. A similar poetic concept was that of the personified north wind, Boreas, also a demigod, winged and wild with disheveled hair and beard, who likewise dwelt in a cave. Why he was supposed to dwell in a cave is not stated, but of course if, as believed, he was real and possessed of human qualities he had to live somewhere, and being superhuman that somewhere could not be like unto the home of a mere mortal, but rather some strange and mysterious place.

Furthermore a cave, and there are many such, that has two openings at widely different levels gives out, on hot days, and from the lower entrance, a strong and continuous cold blast; and surely in the presence of such a weird phenomenon a personificationalist, such as a polytheistic poet, who knew nothing of its true and simple cause, might well attribute it to the blowing by mouth, as we ourselves would blow, of some giant, superhuman being. Perhaps this is why Boreas occasionally was represented, not as winding a conch shell, as by the more aesthetic Greeks, mindful of the noise of the north wind, but with distended cheeks and rounded lips that urge the whistling blast. Similarly, and presumably for much the same reasons, it was common through the middle ages the period of great faith and piety to picture winds as coming from the puckered mouths of wingless cherubs. The well-nigh universal ignorance of winds and their causes is further emphasized by the world-wide belief that by this means or that, and especially by some form or other of imitative magic, one could obtain any kind of wind he might desire, from calm to tempest, and of whatever direction.

Nor were, neither are, these superstitions confined to recognized savages. Ulysses is said to have received the winds in a leathern bag from Aeolus, and quite analogously, for centuries, in certain civilized countries, winds could be purchased, and were purchased, especially by sailors. Perhaps they still may be obtained at a few places, not in leathern bags, as Ulysses got his, for that would be inconvenient, but in knotted strings or handkerchiefs one knot to be untied for a breeze, two for a blow, and three for a storm. But our knowledge of winds has improved of late. We now can talk of the how and the why of them, and discuss their whence and their whither in the exact language of the physicist. We know that air has weight and inertia, and that like mass, or matter, of any other kind a force is required to put it in motion, or to change its speed; and also that when it is not somehow restrained a force acting on it gives it motion, or alters the motion it already has. From this we correctly infer that every wind is caused by a push on the moving air. Furthermore, we know that this push is owing to the unequal distribution of the atmosphere itself over the earth, and that this vastly important irregularity is owing, in turn, mainly to the irregular distribution of temperature.

This, however, is not the whole story, for there is mechanical reaction as well as action. That is, while pressure causes the air to flow the flowing air or wind causes a pressure. That a wind produces a pressure against any obstacle in its path every one will admit. He knows it from the bending, and sometimes breaking, of trees in the blast, and from the push he has felt on his own body when out in a gale.

We also know the magnitude of this push in terms, say, of pounds per square foot of flat surface directly facing the wind, when we know the velocity of the wind and the density of the air. But there also is another and very different force, or reaction, that comes into existence immediately the wind begins to blow, a reaction long unsuspected but of exceeding importance. It comes from the rotation of the earth, which causes an object moving horizontally over the surface continuously to deflect, or at least steadily to push, to the right (assuming one to be facing forward in the direction of travel) in the northern hemisphere, and to the left in the southern hemisphere.

To get some idea as to how this may be, imagine a cannon to be located at the north pole and to be aimed and fired exactly at New York City; let the speed of the projectile be one mile in three seconds, a fair cannon-ball velocity; and let it keep on its course a little above the surface of the earth, and wholly unaffected by it or the winds, until it gets as far south as the original target, or farther if this should be missed.

And it will be missed, for during the time the projectile was on its way from the pole, the earth was continuously rotating from west to east at the rate of one revolution, counted by the fixed stars, in a little less than 24 hours, so that at the end of the projectile's flight of three hours, roughly, necessary for it to get so far south, not New York City, the original target, but some place in California, perhaps San Francisco, will be struck. The projectile therefore in this case has gone or deflected far to the right of its initial course, as viewed from the surface of the earth. If, now, we would keep it directed all the time along the meridian through New York so that the object aimed at would also be hit, and, not missed a few thousand miles, we must arrange a tube, or groove of some kind, to carry the projectile and force it to keep up with the surface of the rotating earth.

In this case the projectile and the western side of the tube or groove would press against each other—a force would be exerted on the projectile to make it keep up with the eastward movement of the earth, and of course an equal westward push would be exerted by the projectile against the groove. This is the deflective "force" of our projectile to the right (in the northern hemisphere) of its path. In reality the projectile merely keeps straight ahead on its course equatorward while the earth at the same time speedily turns under it. Thus the target runs away from the projectile and a wide miss instead of a hit is scored.

Obviously an exactly similar result, a miss, would occur no matter what the target shot at with our super-cannon from the north pole, or south pole either, only in the latter case the deflection would be to the left. Evidently, then, a wind across either pole and directed towards the equator would deflectto the right if from the north pole, to the left if from the south pole—unless prevented from doing so by a suitable pressure against it. All this is obvious enough for a projectile, or a wind, from either pole just mentioned, but what happens at the countless places away from these poles? Exactly the same sort of thing, only with less force, or smaller rate of deviation, because every point on the earth's surface, except those along the equator, is a pole about which the earth is turning with an angular velocity equal to that at the north or south pole times the sine of the latitude of the place.

At the north or south pole it rotates once, measured by the fixed stars, in a little less than 24 hours; about any point 30° from the equator it turns once in just twice that time; and similarly for all other places. Nonsense, we say, and it does seem so. But carefully suspend a good-sized metal ball by a wire from a point some distance above, as Foucault did under the dome of the Pantheon in Paris in 1851, and set it swinging in any direction and watch what happens. If the experiment has been carefully made the pendulum will slowly change the direction of its swing.

We can easily agree that if the pendulum were at the north pole and swinging towards New York City the earth would so turn under it that presently it would be swinging towards Chicago, and a little later towards San Francisco, and so on around until, when the earth had made one complete rotation, it again would be swinging towards New York City.

We can also agree that the pendulum would behave in a similar manner at any place about which as a pole the earth was actually turning and only at such a place. Well, that concession has our logic hemmed in with no possible line of escape, for the Foucault pendulum behaves in just that way wherever it is mounted, save along the equator. Always and everywhere, except on the equator, the direction of its swing slowly changes; that is, in reality, the earth slowly turns around under it, at such rate that the time required to make the complete turn is equal to the time of rotation at the north or south pole divided by the sine of the latitude of the place of observation.

Every place then on the surface of the earth away from the equator is a pole of rotation, and everywhere the earth is rotating with a known angular velocity under the wind at that place. Hence every wind tends to its direction— tends to deflect to the right in the northern hemisphere and to the left in the southern hemisphere, and does thus deflect except in so far as by force it is prevented from doing so. It would seem too that at any particular place this force must be, and it is, the same for one pound of air as for any other pound of air going horizontally with the same speed. That is, this force is proportional to the weight of the object exerting it. It also, though that is not so obvious, is directly proportional to the horizontal velocity of the object moving.

This looks like complication enough, but there still are other forces that must be considered. If we fasten a marble or like object to the end of a string and then, holding the other end fast, whirl it around in a circle we shall find that there is a tension on that string—a pull from each end against the other that increases with the mass of the object whirled, the length of the string and the number of turns per second. Also, if in any way the string is shortened, by allowing it to wrap around one's finger, for instance, or otherwise, the rate of the turning will increase and the pull grow stronger.

Similarly, if water in a large vessel, such as a basin or a bath tub, has some rotary movement and then is allowed to flow out rapidly, the spin in the water will increase as it approaches the outlet until, in many cases, a hollow funnel of air extends from the surface even into the upper end of the drain pipe. Now precisely similar things, and for exactly the same reasons, happen also to the atmosphere. It flows hither and yonder over a rotating earth; rises here under local heating and sinks there owing to local cooling, and there is no rest for it, only ceaseless movement and turmoil. Also, making confusion more confounded, the mountains and valleys, hills, hollows, cliffs, trees and every other roughness on the surface of the earth, stir the winds into myriads of mingling eddies, surges and whirls whose only regularity is their utter irregularity, like a swarm of bees on the wing, only worse, for all the bees are of one size while the eddies are of all sizes.

But really the situation is not that bad provided we are willing to consider one question at a time, and willing also to leave some of the details for the amusement of our successors. As already stated, winds anywhere are the combined result of earth rotation and differences in pressure at the same level. Thus continuous heating of the air in the equatorial regions, such as always obtains, together with its ceaseless cooling in the higher latitudes causes it perpetually to overflow from the former to the latter and to underrun in the opposite direction. This looks simple and obvious, and would be if the earth were not rotating and if the heating and the cooling were continuous and uniformly distributed over their respective areas.

But neither the heating nor the cooling is uniform, and the earth is not still but rotating. However, since the tropical regions are all the time warmer Water Vapor: Distribution and Trends 205 than those near the poles we may assume, as a first rough approximation, that the heating in the former and the cooling in the latter are uniform. This, if the earth were not rotating, would lead to an approximately uniform and continuous overflow from the warmer to the cooler regions and a like return undercurrent from the colder to the warmer zone.

But, as stated, the earth is rotating and that leads to endless complications in the movements of the atmosphere, or winds. Everywhere friction between the surface of the earth and the atmosphere tends to bring the latter to rest with respect to the former—to stop all winds and everywhere produce a dead calm. Therefore air pushed to higher latitudes, where the distance around the earth is less, soon outruns the surface and appears as a west wind. Similarly, air on its way equatorward sooner or later is outrun by the earth and therefore become an east wind. Really this difference in distance around the earth along parallels of unequal latitude is only a part of the cause of east winds in equatorial regions and west winds in middle latitudes. The effect of the additional part, conservation of areas, is even greater than that which would result from the incomplete but obvious cause indicated.

From this it would seem that there should be a belt of winds from the east (east winds) along, and to a considerable distance on each side of, the equator, and beyond this belt another in either hemisphere of west winds. But such aroundthe-world winds would be their own undoing. They come into being owing, as we have seen, to the movement of warm and of cold air to higher and lower latitudes respectively. But such winds, if continuous around the earth would create along and near the interface between them an unbroken belt of high pressure parallel to the equator that would be an impenetrable barrier against any further interchange between the warmer air of the lower latitudes and the colder air of the higher latitudes, whereupon surface friction soon would bring all the air to a dead calm.

As a matter of fact these east winds and west winds, and the belts of high pressure between them, are more or less broken up and irregular. They are well developed and fairly continuous over the Atlantic, Pacific and Indian Oceans, but are only fragmentary and fitful over the continents. In great measure the tropical air makes its first movement towards the polar regions in the form of vast currents to the immediate westward of the ocean links of the high pressure belts; and, in more or less equal measure, the polar air moves towards the equator around the eastern ends of these same links of the high pressure belts.

Trade Winds

The tropical east winds that are over the oceans are called trade winds, in the sense of track winds, or more or less steady winds, along the same course. There are two trade winds on the Atlantic Ocean, namely, the North East Trade and the South East Trade. The first of these comes out of the north around the east end of that link of the northern hemisphere belt of high pressure that crosses the Atlantic Ocean. This wind as it flows southward soon is so outrun by the surface of the rotating earth as to become first a northeast wind and then more and more nearly an east wind.

Its velocity varies from around 6 miles per hour in October to about 12 miles per hour in April. Some of this air ascends, on being decidedly warmed, while some of it crosses the ocean and then turns first northwestward, then northward past now the western end of the same section of the high pressure belt by whose eastern end it had begun in flowing southward. Soon, as it moves northward, it outruns the surface of the earth and turns northeastward to eastward, thus merging with and becoming a part of the prevailing westerlies (winds from the west) of the middle latitudes.

The other trade wind of the Atlantic Ocean comes from the middle latitudes of the southern hemisphere, flowing first northward towards the equator past the eastern end of the Atlantic segment of the southern highpressure belt, then, as the surface of the earth outruns it, turning northwestward and then nearly westward. Its velocity is roughly 12 to 15 miles per hour through the year.

A portion of this South Atlantic trade wind also ascends, owing to surface heating, in tropical regions and flows off at high altitudes towards the adjacent pole, while another portion rounds the western end of the Atlantic segment of the southern high-pressure belt, after which it becomes a portion of the westerlies of the southern hemisphere. Exactly similar trade winds, only feebler as a rule, occur only the Pacific Ocean. Over the Indian Ocean trade winds occur only south of the equator. North of the equator monsoon winds, to be briefly discussed presently, predominate.

Countertrades

So much then for the "trades" which are the steadiest and most persistent of all winds. The next in order of persistence, year in and year out, and which in conjunction with the trades might be of decided advantage in transoceanic aerial travel are the counter trades, or return over-flow branches of the trades, resulting from their ascents in tropical regions. This ascending air flows off towards the higher latitudes of the same hemisphere in which their ascent occurred and after passing latitude 20°, or thereabouts, turn more or less eastward, so that the trade wind beneath and its overhead companion counter trade have very different directions. In this way the counter trades merge with and become portions of the normal or prevailing westerlies.

Westerlies

As already explained, owing to the rotation of the earth, any wind that has come from the general neighbourhood of the equator and attained to latitudes beyond 20° begins to flow eastward, slightly at first and then, with increase of latitude, more and more nearly directly eastward. These winds are the westerlies (winds from westerly points) of middle latitudes. Though this is their average or prevailing direction they are by no means either as regular or persistent as the trades and the countertrades.

Polar Easterlies

Beyond latitude 60° in each hemisphere the surface wind again tends to blow from easterly points, just as it does near the equator. This is because of the flow of the cold air here away from near the pole to places where the distance around the earth along a parallel of latitude is greater, and therefore the eastward speed of the surface correspondingly greater—so much so indeed that the cold air is more or less run away from and thereby rendered an easterly wind. This polar east wind is far better developed over and around Antarctica, where the outflow of the cold air is more pronounced than it is over and around the Arctic Basin where the outflow is comparatively sluggish.

The winds above described, that is, the five trade winds, the five corresponding counter trades, the two westerlies (one in each hemisphere) and the two polar easterlies may be, and often are, called planetary winds, since they have their origin in two general or planetary conditions, namely, (1) rotation, with, in most places, a high surface speed, about an axis that is not far from normal to the direction to the sun, or, as astronomers say, to the plane of the ecliptic; and resulting therefrom, (2) a persistent great difference between the temperatures of the tropical and the polar airs that maintain an equally persistent interzonal circulation.

Monsoons

Another great class of winds, noted for their long duration and constancy of direction, are the monsoons. This word, of Arabic origin, means a season, and is here applied to winds of season-long duration. They occur in many parts of the world but are strongest, most persistent and best known in India and over the adjacent ocean north of the equator. Here, owing to the great contrast in temperature between the continent and the ocean, there is a persistent under or surface flow of air from the land to the water during the colder season, which flow, under the influence of the rotation of the earth, becomes a wind from the northeast—the northeast monsoon.

During the warmer season, on the contrary, the wind is prevailingly from the ocean onto the relatively warm land, and, as before, affected by the rotation of the earth which, in this case, causes the wind to come from the southwest. This is the southwest monsoon that brings the rains to India. The depth of the summer monsoon of India is about 3 miles, and that of the winter monsoon roughly 1.2 miles. Various other places over the earth have greater or less seasonal changes in the directions of their prevailing winds, owing to changes in the temperatures of the adjacent regions, but at no other place are these changes as pronounced, persistent and important as they are in India.

Land Breeze and Sea Breeze

Closely analogous to the monsoon, but vastly less in depth and extent and having only a diurnal period instead of an annual one, are the land breeze and the sea breeze, caused by the difference in temperature between the surface of the water and the surface of the land from the shore back some miles. They are most pronounced where the land is low and arid, skies prevailingly clear and general storm movements only occasional and commonly slight.

The sea breeze, an on-shore daytime wind, is caused by the greater atmospheric pressure, at the time of its occurrence, over the water than over the land, incident to the lower temperature then of the surface of the evaporating water than that of the surface of the immovable soil. At night, especially when there are no clouds to send back radiation to the earth and keep up the temperature of its surface, the temperature contrast is reversed, the water surface then being the warmer of the two. Hence the direction of the horizontal pressure gradient, or push on the air, also is reversed and the surface wind turned back from land to sea. These winds are shallow, roughly 300 to 1500 feet deep, and seldom extend more than 25 miles inland. In most cases they are so graciously refresh- ing as to be as welcome as the family physician or parish priest.

Valley Breeze

Another daytime wind is the so-called valley breeze, a wind, commonly gentle, that on hot and otherwise calm days blows up mountain valleys. It seems to be owing essentially to the heating and expansion of the air over the nearby valley or plain, which heating increases the pressure in the air at a given height, more over the plain than over the adjacent side of the mountain (the air column to the level in question being shorter above the mountain than over the plain) and therefore causes a flow of the air towards and up the mountain sides and valleys.

KATABATIC OR DRAINAGE WINDS

A somewhat analogous class of winds, oppositely directed and ranging from imperceptible to severe, are the mountain breeze, canyon wind, glacier wind, and fall-wind, or, to use their recognized formal name, katabatic winds, of every sort. Another generic name for air movements of this kind is "drainage winds." In all these cases the wind is directed down grade along the surface. In the main it is a night phenomenon and owing to the cooling and consequent contraction and gain in density of the under air, incident partly to its own radiation, but mainly through contact with the relatively cold soil or other substance beneath—cold usually as a result of rapid loss of heat by radiation. In many respects this air drainage is quite analogous to water drainage.

It flows to lower levels, and from uplands runs into gullies, from gullies into drains, from drains into ravines and from ravines into canyons or valleys, and thence, it may be, out onto a plain of lower level. It may flow away gently, or ripplingly like a brook where the grade is moderate, and even plunge over a precipice in the form of an aerial cataract—a cataract, however, that in some cases never reaches the foot of the fall, because in the course of its descent the increasing pressure upon it makes it warmer and therefore lighter than the surrounding air which thereupon quickly arrests its rush to lower levels, and even causes a partial rebound. Drainage winds teach us many a useful lesson.

The mountain camper, for instance, soon learns to pitch his tent above his evening fire, and not below it where the down-flowing air would pester him with the smoke. The Swiss valley dweller sets his house on a knoll so as to be above the nightly flood of cold air that surges around him. The wise orchardist selects for his trees a place part way up a slope, where the air by descent will have appreciably warmed, and yet above the lower levels where the cold air accumulates since it can drain away but slowly.

That is, he plants his orchard along the "thermal belt" where frosts, both early and late, are least likely to occur. The prudent aviator avoids landing at the mouth of a canyon when drainage winds can be strong, because they always are shallow and at times even violent. The cautious boatman when near a steep elevated coast of high latitude keeps his weather eye open for the williwaw, that abrupt, smashing avalanche of frigid air that seems to have sloshed over the mountain brim from the icy bowl beyond. The fisherman of Trieste, Fiume or Novorossiisk, to mention only the most conspicuous of such places, has a wholesome regard for the sudden and chilling Bora of hurricane force that, rushing from the snow-covered uplands pounds the sea into waves and the waves into a cloud of spray. The polar explorer, unless daring the winds to do their worst, as Sir Douglas Mawson seems to have done the year he spent at the "home of the blizzard," keeps away from steep slopes facing the sea.

Of course there are many other examples of gravity winds, as they also are called, or fall winds, another of their various names, but the above are sufficient to emphasize their interest, importance and world-wide distribution—always shallow and ranging from the gentle, balmy drift to a furious, frigid blast of hurricane force.

Tropical Cyclone

A radically different wind from any of those already mentioned, important, often violent and covering a somewhat circular or elliptical area, 50 to 1,000 miles across, is the tropical cyclone—the hurricane of the Atlantic Ocean, Gulf of Mexico, Caribbean Sea and eastern Pacific; baguio of the Philippines; typhoon of the China Sea; and the cyclone of the Indian Ocean. It is characterized by strong winds, up to 100 to 150 miles per hour, directed spirally inward and around a common centre, the place of lowest atmospheric pressure, clockwise in the southern hemisphere, counterclockwise in the northern hemisphere. It is further characterized by dense clouds, much thunder and lightning and torrential rain.

It occurs over all tropical seas and oceans except the South Atlantic, but most frequently over the western North Atlantic, including the Gulf of Mexico and Caribbean Sea, Bay of Bengal and western Pacific, including the China Sea. It starts usually, if not always, in a doldrum or calm region, 5° to 20° away from the equator, and more particularly in that portion which presumably is a very sluggish atmospheric eddy—a section, 100 miles broad, perhaps, of air which though nearly quiet probably has a slight eddy motion owing to the presence on either side of it of oppositely directed "trade" winds. Here the warm air becomes excessively humid, and for that reason, as well as because of its high temperature, comparatively light. Hence convection here becomes pronounced, clouds form and thunderstorms develop.

This ascent of light air necessarily implies the corresponding inflow of surface air, and so it must happen from time to time that, owing to the rotation of the earth, and to the opposition in the directions of the winds on either side of the place of ascent, this incoming air is set into rotation, gently at first, but more and more vigourously as the centre is approached, like disturbed water as it runs into a drain-pipe at the bottom of a basin.

The ascent of the air causes it to cool by expansion, its cooling leads to condensation of some of the vapour present, condensation liberates much heat and this heat in turn continues and enhances the convection—not indefinitely, of course, but to a very considerable extent. Thus once begun this rotating disturbance is likely to perpetuate itself so long as a great quantity of water vapour may be had for condensation. Hence, as the storm is captured by the major neighbouring wind, commonly the trade wind on its poleward side, and often carried thousands of miles across a tropical ocean before reaching land, or being turned off to higher and cooler latitudes, its duration generally is at least several days, often a fortnight, and in extreme cases three weeks or more from its inception in some doldrum region to its last feeble efforts far away where water vapour is scarce and energy from its condensation adequate to the continuation of the storm therefore not available.

Commonly it travels a long way from east to west in a trade wind stream, then slowly turns around the western end of the adjacent link of a high pressure belt and swings off to the east again, but at much higher latitudes and with properties largely different from, and more complicated than, those it formerly possessed. In many cases it is driven onto land where, for a time, it brings heavy rains, and then, as a rule, fades away.

The rate at which these storms use energy—their horsepower—is amazing. Thus from the exceptionally large amount of excellent data available about a particular typhoon the Japanese savant, Horiguti, computed that the latent heat of condensation was freed at a rate equivalent to 1,200,000,000,000 horsepower, and that the rate of its mechanical dissipation of energy was equal to 27,000,000,000 horsepower.

Yet some people talk seriously about destroying the typhoon, or hurricane, by dropping bombs onto it from aeroplanes. How the aviator is to reach his vantage point over such a storm is, of course, his problem. But there is no occasion to worry about the details, for saying "boo," to a blizzard would be in about the same proportion and quite its effective in stopping it.

Extratropical Cyclone

The great contrast in temperature maintained between the polar and equatorial portions of the earth keeps the atmosphere, as previously explained, on a continuous round of circulation between them, because the equality of temperature this circulation tends to produce never is even approximately attained. And because this interzonal circulation is over a rotating earth it follows, as may be demonstrated mathematically, that in mid-latitudes it must occur, and it does so occur, largely in the form of great swirls between broad currents of warm air moving poleward and cold air flowing equatorward. Furthermore, since the warm air moving to higher latitudes outruns the earth it therefore must be to the eastward of its companion cold current flowing equatorward and thereby tending to lag behind the rotating earth.

Hence the swirl thus produced between these two currents, the place where the air tends to run ahead on the eastern side and to lag behind on the western, is a region of atmospheric deficiency, or of low atmospheric pressure, with winds directed round about it counterclockwise in the northern hemisphere, clockwise in the southern, and, in each case, turned spirally inward. The simultaneous directions of the wind all around the centre of this low pressure, or place where the pressure is least, seem, on first considerations, to indicate that the air is spinning spirally around and towards that centre— around and around St. Louis, for instance. But St. Louis is fixed in position, while the centre of the storm is not, but moves on eastward 20 to 30 miles an hour, as a rule, and so one gives up his first inference that St. Louis is the hub of the winds and jumps to the next "obvious" conclusion, namely, that they spiral around and around the traveling centre like the tire of a rolling wheel about the axle.

Well, that is wrong, too, for the tire and the axle do not keep pace with each other. The storm is not a spinning mass of always the same air sliding over the earth, but a traveling swirl of constant type but of ever renewed material. Well then, if both these inferences are wrong, what course does the air follow in such a disturbance? We have learned something about that from the drift of clouds and of free balloons, and from the state and condition of the air at successive times and places, and know now that this course is not by any means so simple as we once thought it was. But we must not let this fascinating flight of an individual air particle develop into a red herring dragged across the course of our subject—wind.

In the area of an extratropical cyclone, roughly elliptical and ranging from 200 to 2000 miles across, there always and everywhere is wind, usually of 10 to 30 miles an hour, but by no means confined within these limits. The cyclone is of frequent occurrence, two or three to a dozen a month, over most of the earth beyond latitudes 30°N, S and most of the winds of these regions are associated with them. As previously explained, the extratropical cyclone is a product of the interzonal exchange of air over the surface of a rotating earth. It therefore consists of two distinct parts, a warm and humid eastern portion where the air is of tropical origin, and a relatively cool and dry western section where the air has come from a higher latitudes. Wherever the boundary, 100 to 1000 miles long, between these two great counter currents of air passes, as it moves over the earth with the storm, there and then the direction of the wind quickly changes, from southwest, likely, to northwest, and its quality from relatively smooth-flowing to gusty. This traveling interface between the two wind systems, the tropical and the polar, is known as the wind-shift line or the polar front. It brings showers of the so-called clearing-up type, many of which in the summer are accompanied by thunder and lightning and squally winds.

A great deal more could be said about the winds of the extratropical cyclone, a disturbance that can be started in three or four different ways, but that would involve tedious details which, except for the specialists, would tend to obscure the major facts here stated and general outlines just drawn.

Anticyclone

The anticyclone, a system of winds over a region where there is more air per unit area, and therefore higher atmospheric pressure, than there is round about, is entirely an extratropical phenomenon, roughly elliptical in shape and commonly 200 to 2000 miles across, which, like the cyclone of the same latitudes, may be originated in more ways than one. Owing to the relatively high pressure within its area the lower air flows outward from an anticyclone, but of course under the influence of the rotation of the earth. Its direction therefore is spirally outward from the centre of highest pressure, and around it, clockwise in the northern hemisphere, counterclockwise in the southern. But as in the case of the extratropical cyclone the direction of the wind in an anticyclone does not mark its course. Near the centre of the anticyclone the winds always are light, while close to its border they are much stronger, sometimes approaching or even reaching gale force. There is, however, a maximum possible velocity which the wind of an anticyclone can have, while no such limit applies to the winds of the cyclone. Another interesting contrast between these major disturbances is the fact that in the cyclone, especially the true cyclone of the tropics and subtropics, the swiftest winds are near the centre of the storm-not at the centre for there the winds often are still, but within 5 to 15 miles of it, while the swiftest winds of the anticyclone are near its border. On the average the velocity of the wind in an anticyclone is only 5 to 15 miles per hour, and distinctly less than that of the cyclone of the same region and season.

One of the most important and striking weather phenomena of the anticyclone is confined to the winter months and is most frequent and severe in the central United States. This is the cold wave that rushing from high latitudes sometimes carries freezing temperatures even to the border of the subtropics. It brings also the dread "norther" of Oklahoma and Texas, and the true blizzard of the Great Plains whose howling, frigid blasts, filled with powdered, blinding snow, give no quarter to man nor beast that ventures forth or is found afield.

Tornado

Child of the cyclone and the anticyclone, of the south wind and the north wind, where brought into closest propinquity, and born among the clouds, is that meteorological enfant terrible, the tornado, the smallest yet most vicious and destructive of all storms. It may be, and generally is, only 1000 feet, or less, in diameter, yet the velocity of its mad, dancing—dervish winds as they wildly spin around and up the central vortex may be 100, 200, even 500 miles an hour. So exceedingly violent is its action that few things can withstand the fury of its blasts. Like one of its parents, the cyclone, its gyration always is counterclockwise in the northern hemisphere and clockwise in the southern, and like its other parent, the anticyclone, it never comes near to the equator. The area covered by this storm is rather small since the length of its path commonly is less than 25 miles. Also its occurrence is, at most, only occasional, hence despite the terror and destruction it works wherever it does hit, the chance of being hit by it at all is so slight that relatively few people, even where it is most likely to occur, show any dread or concern about it.

Waterspout

A usually much smaller and less violent whirling storm than the tornado is the well known waterspout. This is not a rotating column of sea-water, 1000 feet high, as many seem to think, but just a column of cloud, formed in the humid air by its cooling due to the expansion under decrease of pressure incident to rapid rotation. It may be a tornado at sea, in which case it has the same restrictions as to sense, or direction, of rotation, and the same inability to approach the equator. On the other hand it may start from the surface, owing to the local ascent of air that is very humid and therefore relatively light, just as dust whirls are begun by the concentrated ascent of highly heated desert air. Waterspouts of this origin, like their desert brothers, the dust devils, as they often are called, occur at practically all latitudes, even on the equator, and wherever they do occur rotate in either sense, clockwise or counterclockwise. Also in whichever sense a waterspout or a dust whirl starts rotating, that way it continues to rotate as long as it lasts.

The duration of the waterspout commonly is very brief, 15 minutes, or thereabouts. It therefore is an excellent thing with which to prove the efficacy of magic; a little whistling, for instance, or a threat by the pointing of a sword, and it soon vanishes—as it was going to do anyway. Or, if one is more navalminded, he can fire a gun at it—cannon to popgun, it makes no difference. If a gun is not convenient point a stick at it, and if you haven't got a stick your finger will do quite as well, for whatever you do or don't do the spout soon will be only a thing to remember and talk about.

THUNDERSTORM WINDS

Another special class of winds are those that normally accompany the thunderstorm. Commonly this storm is preceded a few minutes by a gentle wind directed towards its nearest portion. This is abruptly followed, just before the rain begins, by a much stronger and distinctly turbulent wind in exactly, or nearly, the opposite direction. The first of these is the indraft of surface air, on its way to great heights where the vapour it carries, or is a part of it, is to condense to cloud and rain and the rain in turn to play its important role in the production of thunder and lightning. The second, or outflowing turbulent wind, has come down from above, owing to increase of density due to cooling, and is swift because it has retained the faster velocity winds usually have at considerable heights.

Eddies

Sometimes one is surprised when, on climbing a mountain with a strong wind at his back, he finds, on going down the opposite side, a wind in his face. He is puzzled to know how the wind can blow up the mountain on both sides at the same time. The mountain does the trick. As the wind crosses the crest it tends to go on at the same level, but it also catches up some of the air beneath it and drags that along, just as the air in the tube of an atomizer is pulled out, and with it some of the liquid beneath, by the blast across its top.

This depletion of the air between the side of the mountain and the sheet of crossing wind, or start towards the production of a vacuum, necessarily leads to an inflow of an equal amount of air from outside, and the most available source is down the mountain. Hence a wind crossing a mountain range induces a return eddy current up the lee slope that is a great puzzle to the uninitiated, and great worry to the student of physics who tries to explain it with all the mathematical and other details satisfactory to an exacting professor.

Foehn

Sometimes the wind that crosses a mountain is so pulled down on the lee side that it strikes the valley or plain beyond, now gently and again even violently. As this crossing air descends to lower levels its temperature increases owing, as we know, and in proportion to, the gain in pressure on it, or amount of air above it. If, now, there is no cloud or rain on the mountain this gain in temperature by the descending air on the lee side is only equal to its cooling in the course of an equal ascent on the windward side, and nothing strikingly out of the ordinary happens.

When, however, abundant rain does fall from the ascending air the cooling is much less, owing to the latent heat of condensation freed, or rendered sensible, than it is when the air is dry, while the amount of warming is substantially unaltered, the air on the lee side of the mountain being clear, or nearly so, in each case. On such occasions the descending air reaches the valley or plain abnormally warm and dry; indeed so great are these effects that now and again they even change fields of deep snow into a dusty plain in a few hours—yes, hours, not days.

TIMES OF GREATEST AND LEAST WINDS

Nearly everywhere over land, but no much so over the ocean, there are diurnal changes of the wind velocity. Over level regions the velocity of the wind is least a little before sun-up, when the surface air is coldest and convection least; that is, when surface friction has acted longest on the same identical air and brought it nearest to rest. Similarly the winds are greatest, on the average, when convection has effectively brought upper, and swiftermoving, air most abundantly down to the surface, commonly 1 to 2 o'clock P.M. On mountain tops the hours of maximum and minimum wind velocity are just the reverse of those of the plain.

Also the winds are stronger in the winter, on the average than in the summer. This is because the push, or horizontal pressure gradient, that makes the winds, is greater, as a rule, in winter than in summer, and this in turn is because the change in temperature with latitude, which largely determines the corresponding change in pressure, is greater in winter than in summer.

Wind Velocity and Height

Another interesting and exceedingly important thing about wind is the Science fact that generally its velocity changes a great deal with change of height. Many studies have been made of this phenomenon, though some of the best of them are pretty well removed from the view of the average individual by, to him, an opaque wall of mathematics. This seclusion was not intentional, but unavoidable owing to the nature of the subject. It is easy to observe what happens, but the details of why it happens are, in many cases, quite difficult to explain.

Whenever there is an appreciable wind the roughnesses on the surface of the earth throw the lower air into a turmoil of rising, falling and mingling eddies. Hence, as the surface is the place where the friction occurs, the velocity of the wind must at first rapidly increase with increase of height, and also progressively change in direction to the extent of 30 or 40 degrees, owing to the rotation of the earth. In many cases this increase of velocity continues, but at a slower rate, up to the level of the highest clouds.

After this the velocity falls off until at twice that level, roughly, it again is no more than at the surface. At still greater heights the velocity once more increases. Often, in the forenoon especially, the increase of wind velocity with height continues up to the 1,500 foot level, or thereabouts, and then decreases until at twice that height the velocity has fallen to surface values. Beyond this level it again increases at a fairly uniform rate, and all the rest of the way behaves as above explained. Occasionally, on the other hand, the air is practically calm from the surface up to great heights—5 or 6 miles at least, and perhaps higher.

❒

4

Biogeochemical Cycle

ECOSYSTEM

An ecosystem is a group of living and non-living components interacting together on a given physical landscape. The size of an ecosystem is arbitrary and could be as small as a few square centimeters if you are looking at a soil microbial ecosystem; as large as thousands of square kilometers if you are looking a biome like the Great Plains ecosystem; or a few hectares if you are looking at a single forest stand ecosystem.

FORESTED ECOSYSTEM WORKS IS TO BUILD A MODEL

An ecosystem model is an accurate but simplified representation of an ecosystem that can be very useful in thinking about or simulating the actions of a real ecosystem.

Because any ecosystem has many different but interrelated components, the best way to understand the system is to break it down into its component parts. To get an introduction to a very simplified forest model, which gives participants and introduction to how a hardwood forest ecosystem works before and after exotic earthworms invade.

Step One: The first step in building a graphical model of a hardwood forest ecosystem is to identify its major components. The components of any ecosystem are those physical things that contain energy and nutrients.

A forested ecosystem, by definition contains trees, so that is our first component. In addition to the various species and layers of trees in a forest, there are other distinct ecosystem components. The understory contains most of the visible plant life found between the sapling layer and forest floor. The forest floor is where one would find most of the plant roots, bulbs, fungi, seeds, years of accumulated leaves and twigs.

The soil is the "dirt" under the forest floor and is composed largely of minerals of various grain size (very small grain size = clay very large grain size = sand) and organic material that has been mixed in with the mineral component. In addition, there are numerous animals that live in the forest, and of course we cannot forget people. We will add those two components to our forest ecosystem model later

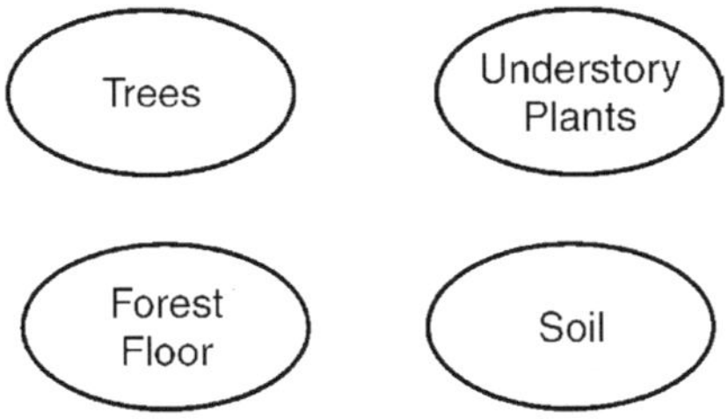

Fig. 4.1: Components of a Graphical Forest Ecosystem

Step Two: Once you have identified the components of your ecosystem model, you need to define the processes that connect the components. This is graphically done by using arrows to indicate the flow of nutrients or energy among the different ecosystem components. The components of our ecosystem model are now connected by processes that result in the movement of energy or nutrients among the components.

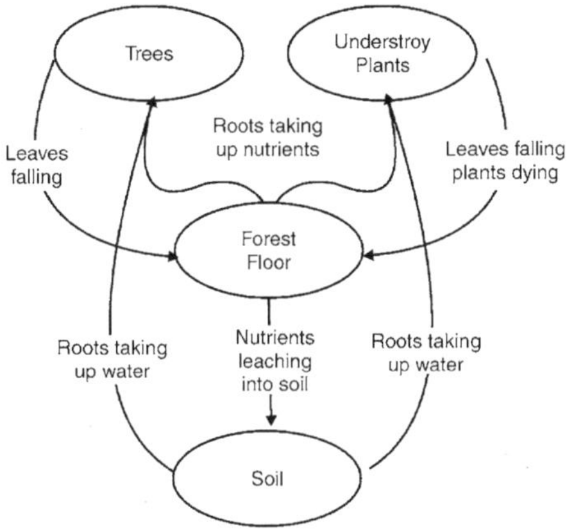

Fig. 4.2: Movement of Energy or Nutrients Among the Components

One thing to notice in our ecosystem model is that there is no process connecting the trees component with the understory plants component. This is because there are no substantial processes that result in the flow of nutrients directly from a tree to an understory plant or visa-versa. There would be important relationships between the trees and understory plants. Trees provide shade to the understory plants. But remember, in an ecosystem model, only processes that result in flow of energy or nutrients are represented. Now let's add the animals and the people components to our ecosystem. energy & nutrients flow from the trees and understory plants to the animals when they eat the leaves, twigs and buds of trees or graze on understory plants; and when the animal excrete waste products or die, energy & nutrients are returned to the forest floor component. Since people are really just a special kind of animal, energy & nutrients flow from the trees to people when they eat something from a tree, like maple syrup

We have added two components to our ecosystem model, along with some processes connecting them to other components.

Step Three: Determine the major inputs and outputs of your ecosystem. As you are building your ecosystem model, one thing to think about is whether your ecosystems could be opened or closed. A closed ecosystem is one that has no inputs of energy or nutrients from outside the ecosystem and no outputs of energy or nutrients leaving the system.

The earth is an example of a closed ecosystem with respect to nutrients and an open ecosystem with respect to energy All the nutrients that have ever been on earth are here and simply continue to cycle, there are no additions or losses. However, the earth is constantly getting inputs of energy from the sun and simultaneously radiating energy back. The earth doesn't heat up too much or cool down too much because the earth's energy balance is in a relatively stable equilibrium, meaning that the amount of energy being input and output are about equal.

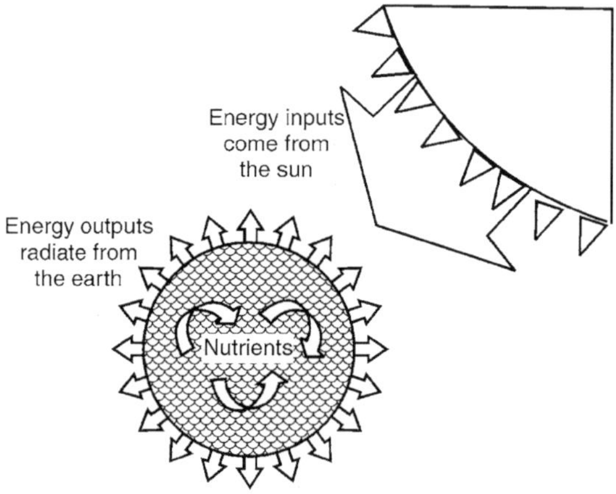

Fig. 4.3: The Earth Ecosystem

The earth ecosystem and has no inputs or outputs of nutrients which are constantly recycled within the global ecosystem, while the earth has both inputs and outputs of energy that are in a relatively stable equilibrium.

Just as the Earth ecosystem is closed with respect to nutrients, unmanaged earthworm-free hardwood forest ecosystems are often very nearly closed nutrient ecosystems that there are very few inputs or outputs of nutrients. Rather the nutrients are constantly recycled among the various ecosystem components. In contrast, most agricultural ecosystems require nutrient inputs from outside to function properly some typical inputs and outputs of nutrients and energy for forested ecosystems include evapotranspiration, nutrient leaching, sunlight and rain.

Step Four: Once you have identified the components, processes and major inputs and outputs in your ecosystem model, then you can begin to add the actual values to these parts of your ecosystem by measuring them. The amount of litter that falls to the forest floor each year (a process), what the biomass of trees is in a given forest (a component), how much light reaches the forest over a growing season (an input), or how much nitrogen leaches from the forest (an output).

Needless to say, some of these things are easier to measure than others and for most of these things it would be very hard to directly measure the value for a whole forest.

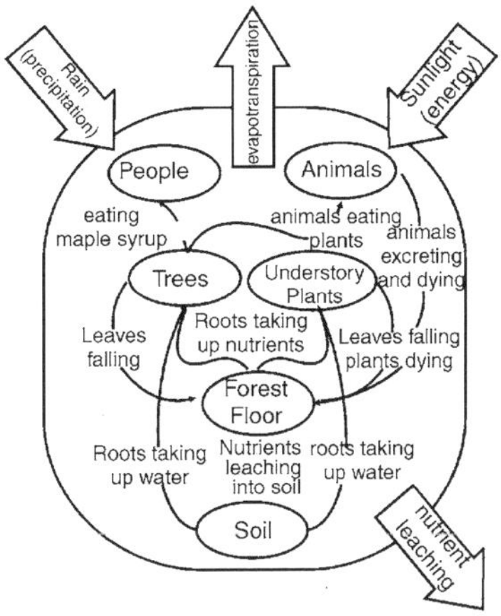

Fig. 4.4: Evapotranspiration

It would be hard to catch every single leaf that fell from the trees in a given year and weigh them all! So, researchers estimate these values taking samples of the given measurement they want to know. In the case of leaf litter, you can put out trays in the forest and after all the leaves have fallen for the year, dry and weight the leaf litre in your trays. They you can use that value to calculate an estimate of the total leaf litter for your forest

Step Five: Use your ecosystem model to think about how changes can cascade through an ecosystem or to ask specific questions that can be answered with further research. When the major components, processes and inputs and outputs of an ecosystem are understood, then you can use the model how changing one part of the ecosystem affects other parts. If you harvest trees from your forest, that will decrease the amount of leaf litter reaching the forest floor each year which may lead to decreases in available nutrients for understory plants. This is the type of thing forest ecology researchers often study.

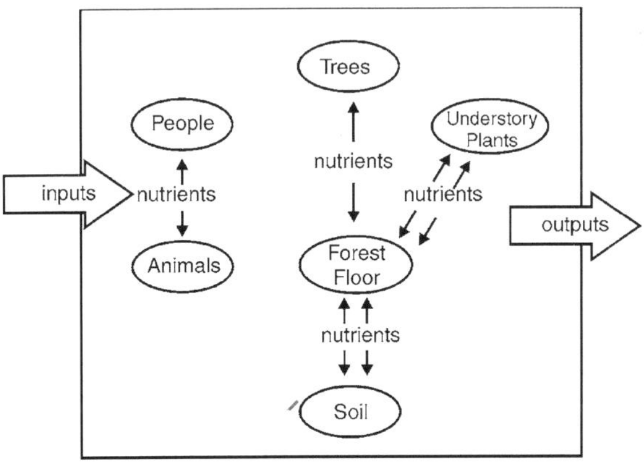

Fig. 4.5: Hard wood forest Ecosystem Model.

An ecosystem that is in equilibrium doesn't gain or lose nutrients. Researchers may monitor soil nutrient levels for many years after trees have been harvested how the real forest behaves compared to what they thought might happen based on their forest model, their understanding of how the forest works.

If the results in the real forest are very different than those predicted by their model, then they know that they don't have full understanding of how their forest works and they may go back to try to improve their model.

In general, when building a model, you start simple and add more detail as needed. To get an introduction to a very simplified forest model which gives participants and introduction to how a hardwood forest ecosystem works before and after exotic earthworms invade.

Energy Flow Through the Ecosystem

Both energy and inorganic nutrients flow through the ecosystem. We need to define some terminology first. Energy "flows" through the ecosystem in the form of carbon-carbon bonds. When respiration occurs, the carbon-carbon bonds are broken and the carbon is combined with oxygen to form carbon dioxide. This process releases the energy, which is either used by the organism or the energy may be lost as heat. The dark arrows represent the movement of this energy.

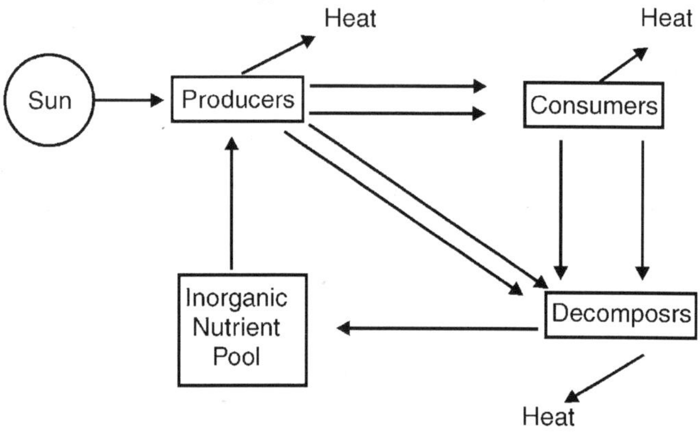

Fig. 4.6: Energy and Inorganic Nutrients flow through the Ecosystem

The other component are the inorganic nutrients. They are inorganic because they do not contain carbon-carbon bonds. These inorganic nutrients include the phosphorous in your teeth, bones, and cellular membranes; the nitrogen in your amino acids; and the iron in your blood.

The movement of the inorganic nutrients is represented by the open arrows. Note that the autotrophs obtain these inorganic nutrients from the inorganic nutrient pool, which is usually the soil or water surrounding the plants or algae.

These inorganic nutrients are passed from organism to organism as one organism is consumed by another. Ultimately, all organisms die and become detritus, food for the decomposers. At this stage, the last of the energy is extracted and the inorganic nutrients are returned to the soil or water to be taken up again. The inorganic nutrients are recycled, the energy is not.

Many of us, when we hear the word "nutrient" immediately think of calories and the carbon-carbon bonds that hold the caloric energy. When writing about energy flow and inorganic nutrient flow in an ecosystem, you must be clear as to what you are referring. Unmodified by "inorganic" or "organic", the word "nutrient" can leave your reader unsure of what you mean. This is one case in which the scientific meaning of a word is very dependent on its context. Another example would be the word "respiration", which to the layperson usually refers to "breathing", but which means "the extraction of energy from carbon-carbon bonds at the cellular level" to most scientists (except those scientists studying breathing, who use respiration in the lay sense).

To summarize: In the flow of energy and inorganic nutrients through the ecosystem, a few generalizations can be made:

- The ultimate source of energy (for most ecosystems) is the sun
- The ultimate fate of energy in ecosystems is for it to be lost as heat.
- Energy and nutrients are passed from organism to organism through the food chain as one organism eats another.
- Decomposers remove the last energy from the remains of organisms.
- Inorganic nutrients are cycled, energy is not.

FOOD CHAINS AND WEBS

A food chain is the path of food from a given final consumer back to a producer. For instance, a typical food chain in a field ecosystem might be:

Grass ⟶ Grasshopper ⟶ Mouse ⟶ Snake ⟶ Hawk

Note that even though I said the food chain is the path of food from a given final consumer back to a producer we typically list a food chain from producer on the left to final consumer on the right. Note to international readers: In Hebrew or Aramaic, or other languages which are read right-toleft, is it customary to list the food chains in the reverse order? By the way, you should be able to look at the food chain above and identify the autotrophs and heterotrophs, and classify each as a herbivore, carnivore, etc. You should also be able to determine that the hawk is a quaternary consumer.

The real world, of course, is more complicated than a simple food chain. While many organisms do specialize in their diets (anteaters come to mind as a specialist), other organisms do not. Hawks don't limit their diets to snakes, snakes eat things other than mice, mice eat grass as well as grasshoppers, and so on. A more realistic depiction of who eats whom is called a food web. It is when we have a picture of a food web in front of us that the definition of food chain makes more sense. We can now a food web consists of interlocking food chains, and that the only way to untangle the chains is to trace back along a given food chain to its source.

The food webs are grazing food chains since at their base are producers which the herbivores then graze on. While grazing food chains are important, in nature they are outnumbered by detritus-based food chains. In detritus-based food chains, decomposers are at the base of the food chain, and sustain the carnivores which feed on them. In terms of the weight (or biomass) of animals in many ecosystems, more of their body mass can be traced back to detritus than to living producers.

Pyramids

The concept of biomass is important. It is a general principle that the further removed a trophic level is from its source (detritus or producer), the less biomass it will contain (biomass here would refer to the combined weight of all the organisms in the trophic level). This reduction in biomass occurs for several reasons:

- Not everything in the lower levels gets eaten
- Not everything that is eaten is digested
- Energy is always being lost as heat

It is important to remember that the decrease in number is best detected in terms or biomass. Numbers of organisms are unreliable in this case because of the great variation in the biomass of individual organisms. For instance, squirrels feed on acorns. The oak trees in a forest will always outnumber the squirrels in terms of combined weight, but there may actually be more squirrels than oak trees. Remember that an individual oak tree is huge, weighing thousands of kilograms, while an individual squirrel weighs perhaps 1 kilogram at best. There are few exceptions to the pyramid of biomass scheme. One occurs in aquatic systems where the algae may be both outnumbered and outweighed by the organisms that feed on the algae. The algae can support the greater biomass of the next trophic level only because they can reproduce as fast as they are eaten. In this way, they are never completely consumed. It is interesting to note that this exception to the rule of the pyramid of biomass also is a partial exception to at least 2 of the 3 reasons for the pyramid of biomass. While not all the algae are consumed, a greater proportion of them are, and while not completely digestible, algae are far more nutritious overall than the average woody plant is (most organisms cannot digest wood and extract energy from it).

Biological Magnification

Biological magnification is the tendency of pollutants to become concentrated in successive trophic levels. Often, this is to the detriment of the organisms in which these materials concentrate, since the pollutants are often toxic.

Biomagnification occurs when organisms at the bottom of the food chain concentrate the material above its concentration in the surrounding soil or water. Producers, as we saw earlier, take in inorganic nutrients from their surroundings. Since a lack of these nutrients can limit the growth of the producer, producers will go to great lengths to obtain the nutrients. They will spend considerable energy to pump them into their bodies. They will even take up more than they need immediately and store it, since they can't be "sure" of when the nutrient will be available again (of course, plants don't think about such things, but, as it turns out, those plants, which, for whatever reason, tended to concentrate inorganic nutrients have done better over the years). The problem comes up when a pollutant, such as DDT or mercury, is present in the environment. Chemically, these pollutants resemble essential inorganic nutrients and are brought into the producer's body and stored "by mistake". This is the first step in biomagnification; the pollutant is at a higher concentration inside the producer than it is in the environment.

The second stage of biomagnification occurs when the producer is eaten. Remember from our discussion of a pyramid of biomass that relatively little energy is available from one trophic level to the next. This means that a consumer (of any level) has to consume a lot of biomass from the lower trophic level. If that biomass contains the pollutant, the pollutant will be taken up in large quantities by the consumer. Pollutants that biomagnify have another characteristic. Not only are they taken up by the producers, but they are absorbed and stored in the bodies of the consumers. This often occurs with pollutants soluble in fat such as DDT or PCB's. These materials are digested from the producer and move into the fat of the consumer. If the consumer is caught and eaten, its fat is digested and the pollutant moves to the fat of the new consumer. In this way, the pollutant builds up in the fatty tissues of the consumers. Water-soluble pollutants usually cannot biomagnify in this way because they would dissolve in the bodily fluids of the consumer. Since every organism loses water to the environment, as the water is lost the pollutant would leave as well. Alas, fat simply does not leave the body.

The "best" example of biomagnification comes from DDT. This long-lived pesticide (insecticide) has improved human health in many countries by killing insects such as mosquitoes that spread disease. On the other hand, DDT is effective in part because it does not break down in the environment. It is picked up by organisms in the environment and incorporated into fat. Even here, it does no real damage in many organisms (including humans). In others, however, DDT interferes with the deposition of calcium in the shells of the bird's eggs. The eggs laid are very soft and easily broken; birds so afflicted are rarely able to raise young and this causes a decline in their numbers. This was so apparent in the early 1960's that it led the scientist Rachel Carson to postulate a "silent spring" without the sound of bird calls. The banning of DDT, the search for pesticides that would not biomagnify, and the birth of the "modern" environmental movement in the 1960's. Birds such as the bald eagle have made comebacks in response to the banning of DDT in the US. Ironically, many of the pesticides which replaced DDT are more dangerous to humans, and, without DDT, disease (primarily in the tropics) claims more human lives.

HUMAN VS. NATURAL FOOD CHAINS

Human civilization is dependent on agriculture. Only with agriculture can a few people feed the rest of the population; the part of the population freed from raising food can then go on to do all the things we associate with civilization. Agriculture means manipulating the environment to favour plant species that we can eat. In essence, humans manipulate competition, allowing favoured species (crops) to thrive and thwarting species which might otherwise crowd them out (weeds). In essence, with agriculture we are creating a very simple ecosystem. At most, it has only three levels-producers (crops), primary consumers (livestock, humans) and secondary consumers (humans). This means that little energy is lost between tropic levels, since there are fewer trophic levels present.

This is good for humans, but what type of "ecosystem" have we created? Agricultural ecosystems have several problems. First, we create monocultures, or fields with only one crop. This is simplest for planting, weeding, and harvesting, but it also packs many similar plants into a small area, creating a situation ideal for disease and insect pests. In natural ecosystems, plants of one species are often scattered. Insects, which often specialize on feeding on a particular plant species, have a hard time finding the scattered plants. Without food, the insect populations are kept in check. In a field of corn however, even the most inept insect can find a new host plant with a jump in any direction. Likewise, disease is more easily spread if the plants are in close proximity. It takes lots of chemicals (pesticides) to keep a monoculture going.

Another problem with human agriculture is that we rely on relatively few plants for food. If the corn and rice crops failed worldwide in the same year, we would be hard-pressed to feed everyone. tural ecosystems usually have alternate sources of food available if one fails.

A final problem associated with agroecosystems is the problem of inorganic nutrient recycling. In a natural ecosystem, when a plant dies it fall to the ground and rots, and its inorganic nutrients are returned to the soil from which they were taken. In human agriculture, however, we harvest the crop, truck it away, and flush it down the toilet to be run off in the rivers to the ocean. Aside from the water pollution problems this causes, it should be obvious to you that the nutrients are not returned to the fields. They have to be replaced with chemical fertilizers, and that means mining, transportation, electricity, etc. Also, the chemical fertilizers tend to run off the fields (along with soil disrupted by cultivation) and further pollute the water.

Some solutions are at hand, but they bring on new problems, too. No-till farming uses herbicides to kill plants in a field; the crop is then planted through the dead plants without plowing up the soil. This reduced soil and fertilizer erosion, but the herbicides themselves may damage ecosystems. In many areas, sewage sludge is returned to fields to act as a fertilizer. This reduces the need for chemical fertilizers, but still requires a lot of energy to haul the sludge around. Further, if one is not careful, things such as household chemicals and heavy metals may contaminate the sewage sludge and biomagnify in the crops which we would then eat.

BIOGEOCHEMICAL CYCLES

We have already seen that while energy does not cycle through an ecosystem, chemicals do. The inorganic nutrients cycle through more than the organisms, however, they also enter into the atmosphere, the oceans, and even rocks. Since these chemicals cycle through both the biological and the geological world, we call the overall cycles biogeochemical cycles. Each chemical has its own unique cycle, but all of the cycles do have some things in common. Reservoirs are those parts of the cycle where the chemical is held in large quantities for long periods of time. In exchange pools, on the other hand, the chemical is held for only a short time.

The length of time a chemical is held in an exchange pool or a reservoir is termed its residence time. The oceans are a reservoir for water, while a cloud is an exchange pool. Water may reside in an ocean for thousands of years, but in a cloud for a few days at best. The biotic community includes all living organisms. This community may serve as an exchange pool (although for some chemicals like carbon, bound in a sequoia for a thousand years, it may seem more like a reservoir), and also serve to move chemicals from one stage of the cycle to another.

For instance, the trees of the tropical rain forest bring water up from the forest floor to be evapourated into the atmosphere. Likewise, coral endosymbionts take carbon from the water and turn it into limestone rock. The energy for most of the transportation of chemicals from one place to another is provided either by the sun or by the heat released from the mantle and core of the Earth. While all inorganic nutrients cycle, we will focus on only 4 of the most important cycles-water, carbon (and oxygen), nitrogen, and phosphorous.

The Water Cycle

In the water cycle, energy is supplied by the sun, which drives evapouration whether it be from ocean surfaces or from treetops. The sun also provides the energy which drives the weather systems which move the water vapour (clouds) from one place to another (otherwise, it would only rain over the oceans). Precipitation occurs when water condenses from a gaseous state in the atmosphere and falls to earth. Evapouration is the reverse process in which liquid water becomes gaseous. Once water condenses, gravity takes over and the water is pulled to the ground.

Gravity continues to operate, either pulling the water underground (groundwater) or across the surface (runoff). In either event, gravity continues to pull water lower and lower until it reaches the oceans (in most cases; the Great Salt Lake, Dead Sea, Caspian Sea, and other such depressions may also serve as the lowest basin into which water can be drawn).

Frozen water may be trapped in cooler regions of the Earth (the poles, glaciers on mountaintops, etc.) as snow or ice, and may remain as such for very long periods of time. Lakes, ponds, and wetlands form where water is temporarily trapped.

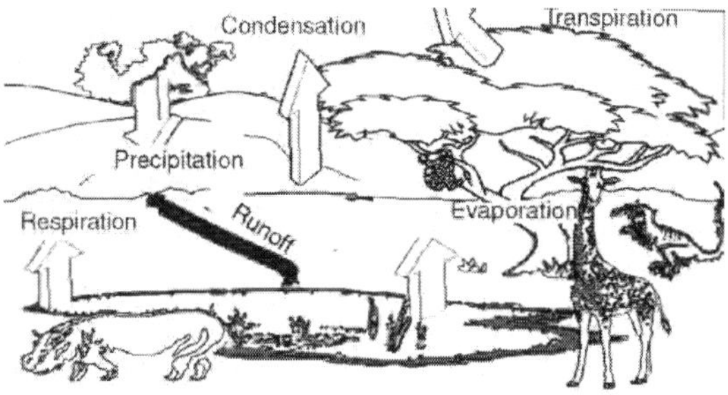

Fig. 4.7: Water Cycle

The oceans are salty because any weathering of minerals that occurs as the water runs to the ocean will add to the mineral content of the water, but water cannot leave the oceans except by evapouration, and evapouration leaves the minerals behind. Thus, rainfall and snowfall are comprised of relatively clean water, with the exception of pollutants (such as acids) picked up as the waster falls through the atmosphere. Organisms play an important role in the water cycle.

As you know, most organisms contain a significant amount of water. This water is not held for any length of time and moves out of the organism rather quickly in most cases. Animals and plants lose water through evapouration from the body surfaces, and through evapouration from the gas exchange structures (such as lungs). In plants, water is drawn in at the roots and moves to the gas exchange organs, the leaves, where it evapourates quickly. This special case is called transpiration because it is responsible for so much of the water that enters the atmosphere. In both plants and animals, the breakdown of carbohydrates (sugars) to produce energy (respiration) produces both carbon dioxide and water as waste products. Photosynthesis reverses this reaction, and water and carbon dioxide are combined to form carbohydrates. Now you understand the relevance of the term carbohydrate; it refers to the combination of carbon and water in the sugars we call carbohydrates.

Carbon Cycle

Once you understand the water cycle, the carbon cycle is relatively simple. From a biological perspective, the key events here are the complementary reactions of respiration and photosynthesis. Respiration takes carbohydrates and oxygen and combines them to produce carbon dioxide, water, and energy. Photosynthesis takes carbon dioxide and water and produces carbohydrates and oxygen. The outputs of respiration are the inputs of photosynthesis, and the outputs of photosynthesis are the inputs of respiration.

The reactions are also complementary in the way they deal with energy. Photosynthesis takes energy from the sun and stores it in the carbon-carbon bonds of carbohydrates; respiration releases that energy. Both plants and animals carry on respiration, but only plants (and other producers) can carry on photosynthesis. The chief reservoirs for carbon dioxide are in the oceans and in rock.

Carbon dioxide dissolves readily in water. Once there, it may precipitate (fall out of solution) as a solid rock known as calcium carbonate (limestone). Corals and algae encourage this reaction and build up limestone reefs in the process. On land and in the water, plants take up carbon dioxide and convert it into carbohydrates through photosynthesis.

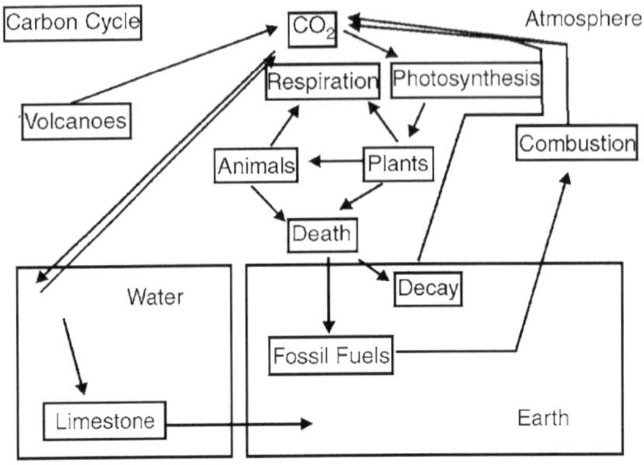

Fig. 4.8: Carbon Cycle

This carbon in the plants now has 3 possible fates. It can be liberated to the atmosphere by the plant through respiration; it can be eaten by an animal, or it can be present in the plant when the plant dies. Animals obtain all their carbon in their food, and, thus, all carbon in biological systems ultimately comes from plants (autotrophs).

In the animal, the carbon also has the same 3 possible fates. Carbon from plants or animals that is released to the atmosphere through respiration will either be taken up by a plant in photosynthesis or dissolved in the oceans. When an animal or a plant dies, 2 things can happen to the carbon in it.

It can either be respired by decomposers (and released to the atmosphere), or it can be buried intact and ultimately form coal, oil, or natural gas (fossil fuels). The fossil fuels can be mined and burned in the future; releasing carbon dioxide to the atmosphere. Otherwise, the carbon in limestone or other sediments can only be released to the atmosphere when they are subducted and brought to volcanoes, or when they are pushed to the surface and slowly weathered away. Humans have a great impact on the carbon cycle because when we burn fossil fuels we release excess carbon dioxide into the atmosphere. This means that more carbon dioxide goes into the oceans, and more is present in the atmosphere. The latter condition causes global warming, because the carbon dioxide in the atmosphere allows more energy to reach the Earth from the sun than it allows to escape from the Earth into space.

The Oxygen Cycle

If you look back at the carbon cycle, we have also described the oxygen cycle, since these atoms often are combined. Oxygen is present in the carbon dioxide, in the carbohydrates, in water, and as a molecule of two oxygen atoms. Oxygen is released to the atmosphere by autotrophs during photosynthesis and taken up by both autotrophs and heterotrophs during respiration.

In fact, all of the oxygen in the atmosphere is biogenic; that is, it was released from water through photosynthesis by autotrophs. It took about 2 billion years for autotrophs (mostly cyanobacteria) to raise the oxygen content of the atmosphere to the 21% that it is today; this opened the door for complex organisms such as multicellular animals, which need a lot of oxygen.

The Nitrogen Cycle

The nitrogen cycle is one of the most difficult of the cycles to learn, simply because there are so many important forms of nitrogen, and because organisms are responsible for each of the interconversions. Remember that nitrogen is critically important in forming the amino portions of the amino acids which in turn form the proteins of your body.

Proteins make up skin and muscle, among other important structural portions of your body, and all enzymes are proteins. Since enzymes carry out almost all of the chemical reactions in your body, it's easy how important nitrogen is. The chief reservoir of nitrogen is the atmosphere, which is about 78% nitrogen.

It is a pretty non-reactive gas; it takes a lot of energy to get nitrogen gas to break up and combine with other things, such as carbon or oxygen. Nitrogen gas can be taken from the atmosphere (fixed) in two basic ways. First, lightning provides enough energy to "burn" the nitrogen and fix it in the form of nitrate, which is a nitrogen with three oxygens attached.

This process is duplicated in fertilizer factories to produce nitrogen fertilizers. The other form of nitrogen fixation is by nitrogen fixing bacteria, who use special enzymes instead of the extreme amount of energy found in lightning to fix nitrogen. These nitrogen-fixing bacteria come in three forms: some are free-living in the soil; some form symbiotic, mutualistic associations with the roots of bean plants and other legumes (rhizobial bacteria); and the third form of nitrogen-fixing bacteria are the photosynthetic cyanobacteria (blue-green algae) which are found most commonly in water. All of these fix nitrogen, either in the form of nitrate or in the form of ammonia (nitrogen with 3 hydrogens attached).

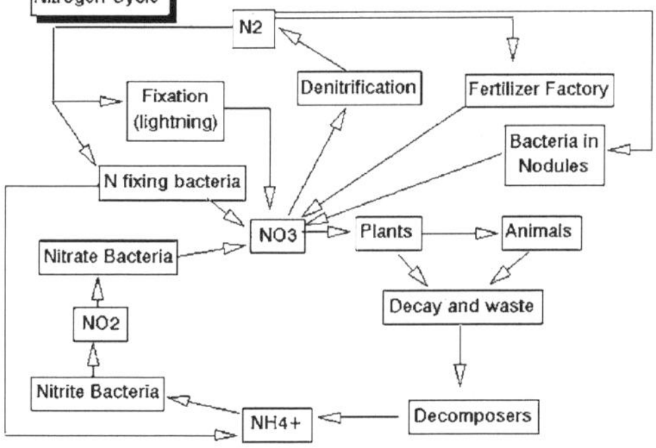

Fig. 4.9: Nitrogen Cycle

Most plants can take up nitrate and convert it to amino acids. Animals acquire all of their amino acids when they eat plants (or other animals). When plants or animals die (or release waste) the nitrogen is returned to the soil. The usual form of nitrogen returned to the soil in animal wastes or in the output of the decomposers, is ammonia. Ammonia is rather toxic, but, fortunately there are nitrite bacteria in the soil and in the water which take up ammonia and convert it to nitrite, which is nitrogen with two oxygens. Nitrite is also somewhat toxic, but another type of bacteria, nitrate bacteria, take nitrite and convert it to nitrate, which can be taken up by plants to continue the cycle. We now have a cycle set up in the soil (or water), but what returns nitrogen to the air? It turns out that there are denitrifying bacteria which take the nitrate and combine the nitrogen back into nitrogen gas.

The nitrogen cycle has some important practical considerations, as anyone who has ever set up a saltwater fish tank has found out. It takes several weeks to set up such a tank, because you must have sufficient numbers of nitrite and nitrate bacteria present to detoxify the ammonia produced by the fish and decomposers in the tank. Otherwise, the ammonia levels in the tank will build up and kill the fish.

This is usually not a problem in freshwater tanks for two reasons. One, the pH in a freshwater tank is at a different level than in a saltwater tank. At the pH of a freshwater tank, ammonia is not as toxic. Second, there are more multicellular plant forms that can grow in freshwater, and these plants remove the ammonia from the water very efficiently. It is hard to get enough plants growing in a saltwater tank to detoxify the water in the same way.

The Phosphorous Cycle

The phosphorous cycle is the simplest of the cycles. Phosphorous has only one form, phosphate, which is a phosphorous atom with 4 oxygen atoms. This heavy molecule never makes its way into the atmosphere, it is always part of an organism, dissolved in water, or in the form of rock.

When rock with phosphate is exposed to water (especially water with a little acid in it), the rock is weathered out and goes into solution. Autotrophs take this phosphorous up and use it in a variety of ways. It is an important constituent of cell membranes, DNA, RNA, and, of course ATP, which, after all, stands for adenosine triphosphate. Heterotrophs (animals) obtain their phosphorous from the plants they eat, although one type of heterotroph, the fungi, excel at taking up phosphorous and may form mutualistic symbiotic relationships with plant roots.

These relationships are called mycorrhizae; the plant gets phosphate from the fungus and gives the fungus sugars in return. Animals, by the way, may also use phosphorous as a component of bones, teeth and shells.

When animals or plants die (or when animals defecate), the phosphate may be returned to the soil or water by the decomposers. There, it can be taken up by another plant and used again. This cycle will occur over and over until at last the phosphorous is lost at the bottom of the deepest parts of the ocean, where it becomes part of the sedimentary rocks forming there. Ultimately, this phosphorous will be released if the rock is brought to the surface and weathered.

Two types of animals play a unique role in the phosphorous cycle. Humans often mine rock rich in phosphorous. For instance, in Florida, which was once sea floor, there are extensive phosphate mines.

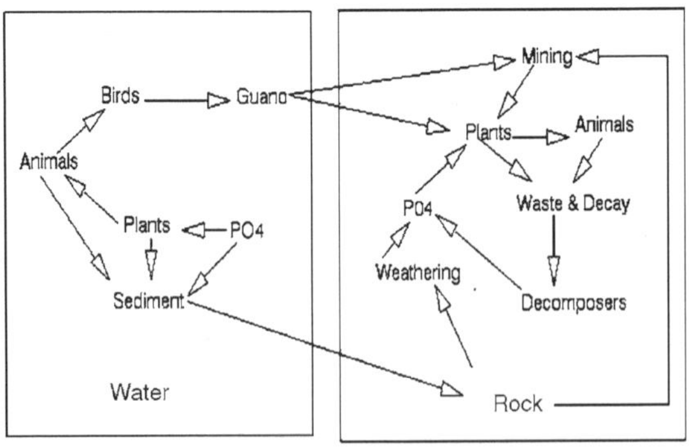

Fig. 4.10: Phosphorous Cycle

The phosphate is then used as fertilizer. This mining of phosphate and use of the phosphate as fertilizer greatly accelerates the phosphorous cycle and may cause local overabundance of phosphorous, particularly in coastal regions, at the mouths of rivers, and anyplace where there is a lot of sewage released into the water (the phosphate placed on crops finds its way into our stomachs and from there to our toilets). Local abundance of phosphate can cause overgrowth of algae in the water; the algae can use up all the oxygen in the water and kill other aquatic life.

This is called eutrophication. The other animals that play a unique role in the phosphorous cycle are marine birds. These birds take phosphorous containing fish out of the ocean and return to land, where they defecate. Their guano contains high levels of phosphorous and in this way marine birds return phosphorous from the ocean to the land. The guano is often mined and may form the basis of the economy in some areas.

Energy Transformations and Biogeochemical Cycling

The study of ecosystems mainly consists of the study of certain processes that link the living, or biotic, components to the non-living, or abiotic, components. Energy transformations and biogeochemical cycling are the main processes that comprise the field of ecosystem ecology. As we learned earlier, ecology generally is defined as the interactions of organisms with one another and with the environment in which they occur. We can study ecology at the level of the individual, the population, the community, and the ecosystem.

Studies of individuals are concerned mostly about physiology, reproduction, development or behaviour, and studies of populations usually focus on the habitat and resource needs of individual species, their group behaviours, population growth, and what limits their abundance or causes extinction. Studies of communities examine how populations of many species interact with one another, such as predators and their prey, or competitors that share common needs or resources.

In ecosystem ecology we put all of this together and, insofar as we can, we try to understand how the system operates as a whole. This means that, rather than worrying mainly about particular species, we try to focus on major functional aspects of the system. These functional aspects include such things as the amount of energy that is produced by photosynthesis, how energy or materials flow along the many steps in a food chain, or what controls the rate of decomposition of materials or the rate at which nutrients are recycled in the system.

Components of an Ecosystem

A basic understanding of the diversity of plants and animals, and how plants and animals and microbes obtain water, nutrients, and food. We can clarify the parts of an ecosystem by listing them under the headings "abiotic" and "biotic".

Abiotic Components	Biotic Components
Sunlight	Primary producers
Temperature	Herbivores
Precipitation	Carnivores
Water or moisture	Omnivores
Soil or water chemistry (e.g., P, NH_4^+)	Detritivores
All of these vary over space/time	

By and large, this set of environmental factors is important almost everywhere, in all ecosystems. A functional group is a biological category composed of organisms that perform mostly the same kind of function in the system; all the photosynthetic plants or primary producers form a functional group. Membership in the functional group does not depend very much on who the actual players happen to be, only on what function they perform in the ecosystem.

PROCESSES OF ECOSYSTEMS

This figure with the plants, zebra, lion, and so forth illustrates the two main ideas about how ecosystems function: ecosystems have energy flows and ecosystems cycle materials.

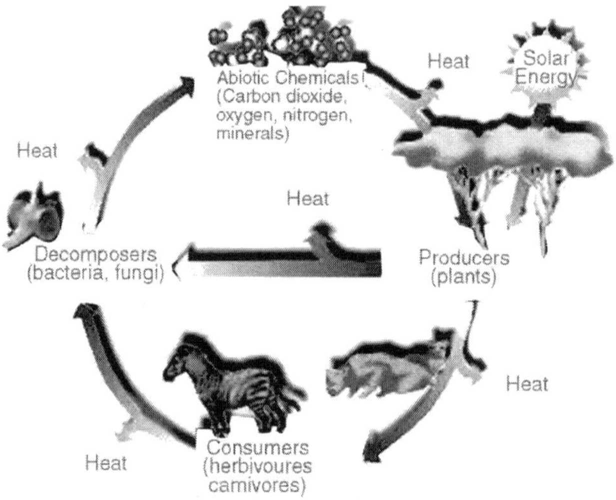

Fig. 4.11: Energy Flows and Material Cycles.

These two processes are linked, but they are not quite the same energy enters the biological system as light energy, or photons, is transformed into chemical energy in organic molecules by cellular processes including photosynthesis and respiration, and ultimately is converted to heat energy. This energy is dissipated, meaning it is lost to the system as heat; once it is lost it cannot be recycled.

Without the continued input of solar energy, biological systems would quickly shut down. Thus the earth is an open system with respect to energy. Elements such as carbon, nitrogen, or phosphorus enter living organisms in a variety of ways. Plants obtain elements from the surrounding atmosphere, water, or soils. Animals may also obtain elements directly from the physical environment, but usually they obtain these mainly as a consequence of consuming other organisms.

These materials are transformed biochemically within the bodies of organisms, but sooner or later, due to excretion or decomposition, they are returned to an inorganic state. Often bacteria complete this process, through the process called decomposition or mineralization During decomposition these materials are not destroyed or lost, so the earth is a closed system with respect to elements (with the exception of a meteorite entering the system now and then). The elements are cycled endlessly between their biotic and abiotic states within ecosystems. Those elements whose supply tends to limit biological activity are called nutrients.

The Transformation of Energy

The transformations of energy in an ecosystem begin first with the input of energy from the sun. Energy from the sun is captured by the process of photosynthesis. Carbon dioxide is combined with hydrogen (derived from the splitting of water molecules) to produce carbohydrates (CHO). Energy is stored in the high energy bonds of adenosine triphosphate.

The prophet Isaah said "all flesh is grass", earning him the title of first ecologist, because virtually all energy available to organisms originates in plants. Because it is the first step in the production of energy for living things, it is called primary production Herbivores obtain their energy by consuming plants or plant products, carnivores eat herbivores, and detritivores consume the droppings and carcasses of us all. portrays a simple food chain, in which energy from the sun, captured by plant photosynthesis, flows from trophic level to trophic level via the food chain.

A trophic level is composed of organisms that make a living in the same way, that is they are all primary producers (plants), primary consumers (herbivores) or secondary consumers (carnivores). Dead tissue and waste products are produced at all levels.

Scavengers, detritivores, and decomposers collectively account for the use of all such "waste"—consumers of carcasses and fallen leaves may be other animals, such as crows and beetles, but ultimately it is the microbes that finish the job of decomposition. Not surprisingly, the amount of primary production varies a great deal from place to place, due to differences in the amount of solar radiation and the availability of nutrients and water.

Energy transfer through the food chain is inefficient. This means that less energy is available at the herbivore level than at the primary producer level, less yet at the carnivore level, and so on. The result is a pyramid of energy, with important implications for understanding the quantity of life that can be supported. Usually when we think of food chains we visualize green plants, herbivores, and so on. These are referred to as grazer food chains, because living plants are directly consumed. In many circumstances the principal energy input is not green plants but dead organic matter.

These are called detritus food chains. Examples include the forest floor or a woodland stream in a forested area, a salt marsh, and most obviously, the ocean floor in very deep areas where all sunlight is extinguished 1000's of meters above. In subsequent lectures we shall return to these important issues concerning energy flow. Finally, although we have been talking about food chains, in reality the organization of biological systems is much more complicated than can be represented by a simple "chain". There are many food links and chains in an ecosystem, and we refer to all of these linkages as a food web. Food webs can be very complicated, where it appears that "everything is connected to everything else", and it is important to understand what are the most important linkages in any particular food web.

Biogeochemistry

One obvious way is to study the flow of energy or the cycling of elements. The cycling of elements is controlled in part by organisms, which store or transform elements, and in part by the chemistry and geology of the natural world. The term Biogeochemistry is defined as the study of how living systems influence, and are controlled by, the geology and chemistry of the earth. Thus biogeochemistry encompasses many aspects of the abiotic and biotic world that we live in.

There are several main principles and tools that biogeochemists use to study earth systems. Most of the major environmental problems that we face in our world toady can be analysed using biogeochemical principles and tools. The principles and tools that we use can be broken down into 3 major components: element ratios, mass balance, and element cycling.

Element Ratios

In biological systems, we refer to important elements as "conservative". These elements are often nutrients. By "conservative" we mean that an organism can change only slightly the amount of these elements in their tissues if they are to remain in good health. It is easiest to think of these conservative elements in relation to other important elements in the organism. In healthy algae the elements C, N, P, and Fe have the following ratio, called the Redfield ratio after the oceanographer who discovered it:

$$C: N: P: Fe = 106: 16: 1: 0.01$$

Once we know these ratios, we can compare them to the ratios that we measure in a sample of algae to determine if the algae are lacking in one of these limiting nutrients.

Mass Balance

Another important tool that biogeochemists use is a simple mass balance 130 Encyclopaedia of Environmental Science equation to describe the state of a system. The system could be a snake, a tree, a lake, or the entire globe. Using a mass balance approach we can determine whether the system is changing and how fast it is changing. The equation is:

$$\text{Net Change} = \text{Input} + \text{Output} + \text{Internal Change}$$

In this equation the net change in the system from one time period to another is determined by what the inputs are, what the outputs are, and what the internal change in the system was. The example given in class is of the acidification of a lake, considering the inputs and outputs and internal change of acid in the lake.

Element Cycling

Element cycling describes where and how fast elements move in a system. There are two general classes of systems that we can analyse. A closed system refers to a system where the inputs and outputs are negligible compared to the internal changes. Examples of such systems would include a bottle, or our entire globe. There are two ways we can describe the cycling of materials within this closed system, either by looking at the rate of movement or at the pathways of movement.

- Rate = number of cycles/time * as rate increases, productivity increases
- Pathways-important because of different reactions that may occur In an open system there are inputs and outputs as well as the internal cycling. Thus we can describe the rates of movement and the pathways, just as we did for the closed system, but we can also define a new concept called the residence time. The residence time indicates how long on average an element remains within the system before leaving the system.
- Rate
- Pathways
- Residence time, Rt Rt = Total amount of matter/output rate of matter

CONTROLS ON ECOSYSTEM FUNCTION

Now that we have learned something about how ecosystems are put together and how materials and energy flow through ecosystems, we can better address the question of "what controls ecosystem function"? There are two dominant theories of the control of ecosystems. The first, called bottom-up control, states that it is the nutrient supply to the primary producers that ultimately controls how ecosystems function.

If the nutrient supply is increased, the resulting increase in production of autotrophs is propagated through the food web and all of the other trophic levels will respond to the increased availability of food (energy and materials will cycle faster). The second theory, called top-down control, states that predation and grazing by higher trophic levels on lower trophic levels ultimately controls ecosystem function.

If you have an increase in predators, that increase will result in fewer grazers, and that decrease in grazers will result in turn in more primary producers because fewer of them are being eaten by the grazers. Thus the control of population numbers and overall productivity "cascades" from the top levels of the food chain down to the bottom trophic levels. So, which theory is correct? Well, as is often the case when there is a clear dichotomy to choose from, the answer lies somewhere in the middle.

There is evidence from many ecosystem studies that both controls are operating to some degree, but that neither control is complete. The "top-down" effect is often very strong at trophic levels near to the top predators, but the control weakens as you move further down the food chain. Similarly, the "bottom-up" effect of adding nutrients usually stimulates primary production, but the stimulation of secondary production further up the food chain is less strong or is absent. Thus we find that both of these controls are operating in any system at any time, and we must understand the relative importance of each control in order to help us to predict how an ecosystem will behave or change under different circumstances, such as in the face of a changing climate.

THE GEOGRAPHY OF ECOSYSTEMS

There are many different ecosystems: rain forests and tundra, coral reefs and ponds, grasslands and deserts. Climate differences from place to place largely determine the types of ecosystems. How terrestrial ecosystems appear to us is influenced mainly by the dominant vegetation. The word "biome" is used to describe a major vegetation type such as tropical rain forest, grassland, tundra, etc., extending over a large geographic area. It is never used for aquatic systems, such as ponds or coral reefs. It always refers to a vegetation category that is dominant over a very large geographic scale, and so is somewhat broader than an ecosystem. Every place on earth gets the same total number of hours of sunlight each year, but not the same amount of heat.

The sun's rays strike low latitudes directly but high latitudes obliquely. This uneven distribution of heat sets up not just temperature differences, but global wind and ocean currents that in turn have a great deal to do with where rainfall occurs. Add in the cooling effects of elevation and the effects of land masses on temperature and rainfall, and we get a complicated global pattern of climate.

A schematic view of the earth shows that, complicated though climate may be, many aspects are predictable High solar energy striking near the equator ensures nearly constant high temperatures and high rates of evapouration and plant transpiration. Warm air rises, cools, and sheds its moisture, creating just the conditions for a tropical rain forest. Contrast the stable temperature but varying rainfall of a site in Panama with the relatively constant precipitation but seasonally changing temperature of a site in New York State. Every location has a rainfall- temperature graph that is typical of a broader region.

We can draw upon plant physiology to know that certain plants are distinctive of certain climates, creating the vegetation appearance that we call biomes. Note how well the distribution of biomes plots on the distribution of climates Note also that some climates are impossible, at least on our planet. High precipitation is not possible at low temperatures—there is not enough solar energy to power the water cycle, and most water is frozen and thus biologically unavailable throughout the year. The high tundra is as much a desert as is the Sahara.

- Ecosystems are made up of abiotic (non-living, environmental) and biotic components, and these basic components are important to nearly all types of ecosystems. Ecosystem Ecology looks at energy transformations and biogeochemical cycling within ecosystems.
- Energy is continually input into an ecosystem in the form of light energy, and some energy is lost with each transfer to a higher trophic level. Nutrients, on the other hand, are recycled within an ecosystem, and their supply normally limits biological activity. So, "energy flows, elements cycle".
- Energy is moved through an ecosystem via a food web, which is made up of interlocking food chains. Energy is first captured by photosynthesis (primary production). The amount of primary production determines the amount of energy available to higher trophic levels.
- The study of how chemical elements cycle through an ecosystem is termed biogeochemistry. A biogeochemical cycle can be expressed as a set of stores (pools) and transfers, and can be studied using the concepts of "stoichiometry", "mass balance", and "residence time".
- Ecosystem function is controlled mainly by two processes, "topdown" and "bottom-up" controls.
- A biome is a major vegetation type extending over a large area. Biome distributions are determined largely by temperature and precipitation patterns on the Earth's surface.

BIOGEOCHEMISTRY

As the terminology implies, Biogeochemistry is a borderline specialisation involving three basic branches of natural, physical and earth sciences. On one hand living organisms are involved and on the other substances such as nutrients, are made available to the living organisms either through soil (the pedosphere) or through water (the hydrosphere). Of course there are certain organisms which can get their food through air (atmosphere) also. The living organisms, while differing vastly among themselves in size, morphology and physiology, share, all the same a basic feature common to them all: ability to metabolise their requirements in the habitats they live. An individual organism is perishable but life process, sustained by successive generation, is perpetual. Biogeochemistry is all about impact of such living processes on our environment. The bulk of living matter is concentrated in the landmass of the world though at the continent-ocean interface and at the bottom of the sea they are present in abundance. The physio-chemical interaction between organisms and the land that sustain them have been a subject of study for over a century now.

More than two hundred years ago, The turnover of elements on the surface of the globe in which he postulated on the cyclic exchange of elements among the three kingdoms of nature, namely mineral, vegetable and animal. Lavoisier paid a heavy price for stating so: he was executed after his work was published. The interaction among the above three kingdoms essentially constitute the broad field of Biogeochemistry as we know today. The original kingdoms have been replaced by reservoirs of different kind and atmosphere is brought into the picture because of gases involved in many biogeochemical processes. Since life is cyclic, all processes where living matter is involved has also got to be cyclic. Hence, in all the reservoirs are linked to each other in both directions:at the inlet and outlet point for each reservoir. The interconnection of all the reservoirs leads to a overall global cycle.

Such a cycle is known by various names depending on the user and the sector of interest: biogeochemical cycle, hydrological cycle, geochemical cycle and rock cycle. The emphasis in each cycle varies: in hydrological cycle, one will be interested in transfer of water; in the geochemical cycle, one will be interested in transfer from the crust to other segments and in rock cycle, the interest will be transformation of large volume of crustal materials across various interfaces. Whereas in the biogeochemical cycle the focus is on nutrients essential for the life processes and at the toxic substances determinant to such living things. Thus biogeochemical processes can be defined in three stages:

Stage-1: Elements to be grouped according to the needs:

Essential, such as Carbon (C), Nitrogen (N) and Phosphorous (P)

Minor, such as Sodium (Na), Potassium(K) etc.

Trace, such as Iron (Fe), Manganese (Mn), Copper (Cu) etc.

Toxic, such as Mercury (Hg), Selenium (Se), Cadmium (Cd) (all the above groupings according to the requirements of biological system in the biological world)-the bio part of the topic.

Stage-2: The availability of the above from the earth, generally from the top few metres of the earth (the geocomponent, more specifically the pedosphere the geo part of the topic and now.

Stage-3: The physio-chemical mechanisms that make the needs of the biological systems from the pedospheric layers of the earth in response to micro-environmental functions such as pH (acidity), weathering and release of elements by their breakdown due to chemical interaction of individual minerals with water. This constitutes the chemical part of the topic.

Let this linkages using an essential element Carbon as an example. Referring to carbon is present in the atmosphere mainly as carbon dioxide (CO_2) and in small quantities as carbon monoxides (CO). In the hydrosphere, carbon is present as bicarbonate and carbonate the major factor in the hardness of water. In the lithosphere- pedosphere, it is present as calcium carbonate ($CaCO_3$) or some combination thereof. In the biosphere, carbon is present as the building block of the plants and animals in the form of variety of carbon compounds such as carbohydrates, proteins and amino-acids. Carbon enters the biological system either through geological process (weathering of carbonbearing minerals such as limestone ($CaCO_3$) and /or chemical process directly from the atmosphere (CO_2) by photosynthesis.

When the organism (including plant) dies out, the carbon is released back either directly to the atmosphere as CO_2 or is mineralised as coal, limestone etc. So that it is returned to the lithosphere. Thus the carbon cycle is complete. The oldest carbon-bearing organic world remains known to man is about 3.5 billion years whereas the age of the earth is at present estimated to be about 4.5 billion years. In spite of the complexity of the biological evolution through geological time scale, carbon has been and is being cycled through various reservoirs relatively unaffected and un- disturbed till recent times.

Among the three essential nutrients for life phosphorous (P) is extremely interesting. Unlike carbon and nitrogen, it is predominately derived from the crustal top of the solid earth and mobilised through the biosphere by the transport behaviour in the hydrosphereP ignites in air but glows when left in darkness. Thus it is a carrier of energy in all living systems. The homologous elements P and N has complimentary properties in relation to Hydrogen(H) and Oxygen(O_2)- the other important elements for most living matter on the surface of the earth. P in association with O_2 introduces some structural order in the molecules at cellular levels and thus the P-O association in the form of PO_4 units separates life from ordinary inanimate minerals represented by similar association of Silicon with O2 (SiO_4)

From a single cell amoeba some 3.5 billion years ago (3500 million), organisms progressively went through complex biological evolution as a function of geological time scale. The link between time and evolution is established through changing environmental parameters such the carbon dioxide (CO_2) levels in atmosphere prevailing at a given time, the O_2 levels, nutrients such as P and minor and trace elements etc. Geochemical evidences through rock records show that about 2.5 billion years ago, the atmospheric CO_2 was at least 100 times lower as that of today and O_2 levels negligible compared to today s environment. have been continuously interacting among each other through such a vast time scale to maintain the cyclic behaviour of nutrients and other elements. Excessive constituents in any one segment was removed over a period of time to other segments so that its biogeochemical cyclic behaviour was not affected. During Gondwana times (about 400 to 500 million years before) when India, Australia and Antarctica were neighbours, excess CO_2 in the atmosphere was neutralised by enhanced photosynthesis and subsequently the biota decayed and gave raise to most of the coal deposits which we are exploiting today.

Thus due to the biogeochemical behaviour, nature was able to adjust itself at its own rate in any parameters. On the other hand in modern times, man introduces any one variable to a segment in very abruptly and in such large dose that nature is unable to adjust to that change. A case in point the emission of green house gas CO_2 by the burning of coal. The essential nutrient element C is transferred from the lithosphere to the atmosphere so fast that the excess CO_2 is unable to be removed equally faster by the hydrosphere leading to enhanced levels of CO_2 in the atmosphere and hence we hear everyday about the global warming On the other hand, let us take a element such as Mercury (Hg) considered to be very toxic to the biological world. The movement of Hg in cyclic fashion through our environment. Hg generally occurs in minerals in association with sulphur in the form of mercury sulphide (HgS) systems.

Even in rocks, it is considered to be trace element since its abundance in the crustal earth is very low. However, during the exploitation of base metals resources, such as that of copper, lead and Zinc, mercury is extracted as a buy-product. Modern civilization uses mercury for a variety of products such as chemicals, coatings on seeds etc. Even though it does not react with water in the hydrosphere easily, mercury combines with any organic substances such decayed plant debris etc and immediately becomes toxic to the living biota. The well known minemata decease in Japan was the result of such affinity of mercury for many biotic materials. Being the only metal known to man to be in liquid state at room conditions, it easily vapourises and hence get also into the atmospheric segment Hence due to man s use, what was once locked up safely in inert geological materials in rocks, have been mobilised to the biosphere, hydrosphere and atmosphere.

The story is basically same for a few other toxic metals such as Cadmium though the details may individually vary due to their differing biogeochemical behaviour in our present environment.

It is clear that processes in our environment are better understood by focussing on the interfacing areas. Thus Biogeochemistry is the ideal such frontier area of study and research. It is however a relatively new field and only in the last two decades various researchers across the world have been focussing their attention on this inter- disciplinary subject. Outside India there are a few places where separate institutes exists for this purpose. At about the same time the Ministry of Environment and Forests also approved the setting up of a ENVIS centre in JNU on the same subject under his direction. Though there no other specific institutions in India in this subject, there are individual researchers in many universities working on topics of interest to Biogeochemistry.

Thus, the area of Biogeochemistry is an emerging branch of multidiciplinary approach involving all major branches of Sciences. From a humble beginning two decades ago, now this subject is well established in the international field and in India is fast catching up as an important tool of research for understanding processes in our environment. Sufficient infrastructure has now developed in India through a variety of individual and institutional efforts so that in the years to come the importance work on Biogeochemistry will be better understood and appreciated.

The involvement of a number of specialists is obvious when ones looks into the above three stages: for understanding the biotic behaviour, one should have a good understanding of ecological principles; the requirements for the ecosystem is provided by basic geochemical processes of weathering helped by microbiological and biochemical reactions in soils and over rocks; farther, from the source to the receptor, chemical and hydrological principles operate as a middleman. Hence within the broad field of Biogeochemistry itself, there are sub-specialisations depending on the background of the researchers- be a ecologist, geochemist or chemical hydrologist. Hence the word Biogeochemistry is truly an integrating field to better understand our present, past and future environmental problems.

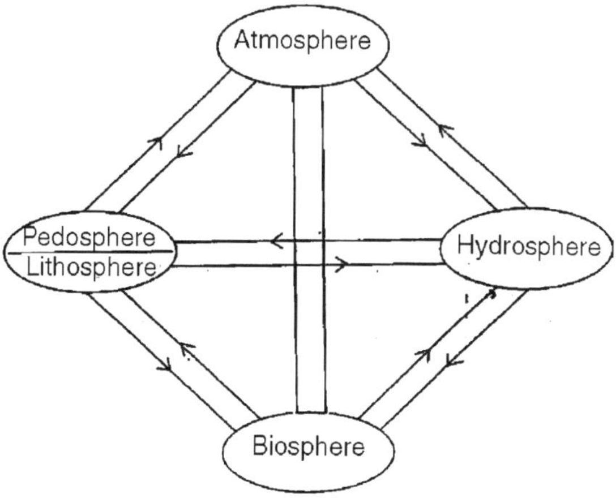

Fig. 4.12: Receptors and Donors to other Neighbouring Spheres

Among the more recent sub-specialisations within Biogeochemistry that have emerged in the international research is the application of certain biochemical and molecular biochemical techniques to understand our present and past environment. A case in point is the study about the Alpine-man.

A full human body was found in tact in the frozen alps and several investigations on the body using most modern biochemical techniques have thrown some light about the habitat of the community some 5000 years ago, then the prevailing eco- system, palaeo-environment, anthropological lineage etc. DNA based study thus offers excellent scope even in environment related matter. In India, a student from our labouratory is working under the supervision of Prof. Laljit Singh, Centre for DNA Finger-printing at Hyderabad to apply DNA techniques in fossils bones of geological and anthropological importance to understand the habitat and palaeo-environment. DNA techniques to ancient biological specimens. Farther, characterisations of aminoacids in ancient sediments have been successfully used to understand palaeoenvironment and also stresses acting on such systems. In India, however, this has not been attempted so far.

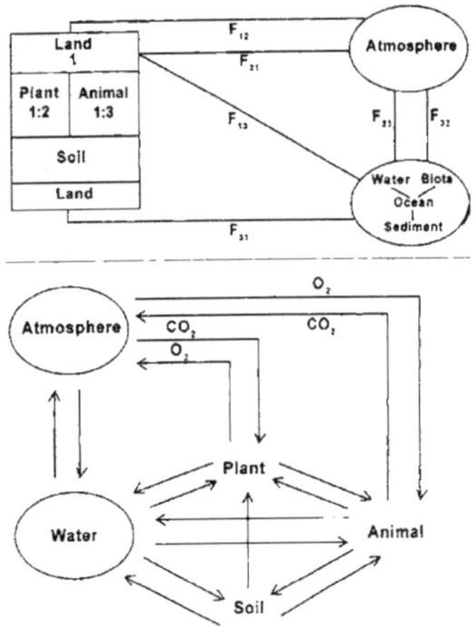

Fig. 4.13: Interaction in the Biogeochemical Cycle

Thus, Biogeochemistry can be viewed as a very challenging area of study involving a variety of basic sciences so that it is ideally suited to understand our environment better. Research should not go in the direction of reinventing the wheel but improve and if possible, replace it. Hopefully the 21st century consolidation of the field of Biogeochemistry as an established brach of science. Enlarged version of specific component. whenevr possible. Important components in each segment and direction of transfer F means flux that is concentration for a given parameter the volume of water/air transferred in that direction per unit time. This, F has unit of wt/vol/time. F is likely to have volumes since the time involved is relatively shorter for some of the sectors. For eg. some of the variables transferred involving soil and plants/animals and plants and atmosphere etc. may have a time scale of a few seconds only.

The carbon transfer, the following fluxes have been incorporated: 1 to 34 point all have same magnitude, time scale and importance. Between river and young soils sediments, the transfer 26 and 27 represent deposition and erosion processes. Faster compared to deposition or accumulation so that net transfer is loss of carbon from soil (Pedosphere) to river (hydrosphere). Similarly, due to deforestation, evapouration may be higher than the net river volume.

Hg cycle is shwn in tow stages: pre-man and present day. this concept is adapted from Garrels et al., 1976. The main differences is indicated in teh Hgmining in the present day cause additional flux-vaporisation-from land to atmosphere. ++ or + singns indicate higher gains in the respectinve sectors the present day cycle relative to pre-man cycle. As a consequence there is also poportional increase in the flxes in the present day cycle between the concerned sectors. Garels et al. (1976) estimate that Hg will be the first metal that is likely to be lost out entiely ot of the solid spheres (litho-and pedospheres) to the fluid (hydro- and atmo-) spheres in foreseabe future.

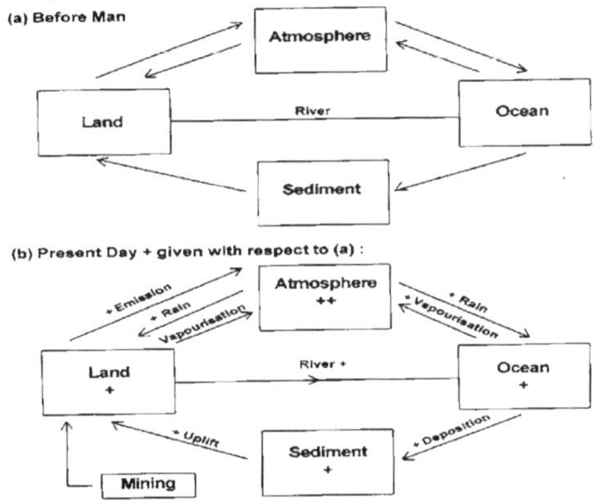

Fig. 4.14: Hg in Biogeochemical Cycle

Mercury in the Environment

Mercury is a silvery, liquid metal at room temperature and is often referred to as one of the "heavy metals." Like water, mercury can evapourate and become airborne. Because it is an element, mercury does not break down into less toxic substances. Once mercury escapes to the environment, it circulates in and out of the atmosphere until it ends up in the bottoms of lakes and oceans. Mercury can be found as the elemental metal or in a wide variety of organic and inorganic compounds. Depending on its chemical form, mercury may travel long distances before it falls to earth with precipitation or dust. Bacteria and chemical reactions in lakes and wetlands can change the mercury into a much more toxic form known as methylmercury. Fish become contaminated with methylmercury by eating food (plankton and smaller fish), which has absorbed methylmercury.

As long as the fish continue to be exposed to mercury, mercury continually builds up in fish's bodies. Fish that eat other fish become even more highly contaminated. Thus, the largest tend to be the most contaminated. When people eat the contaminated fish, the methylmercury can remain in their bodies for a long time. If they eat fish containing methylmercury faster than their bodies can get discharge it, the methylmercury accumulates in their bodies and can be toxic. Many states have fish consumption advisories to inform people about how many meals of fish they can safely eat over a period of time.

Mercury's Environmental Effects

Fish are the main source of food for many birds and other animals, and mercury can seriously damage the health of these species. Loons, eagles, panthers, otters, mink, kingfishers and ospreys naturally eat large quantities of fish. Because these predators rely on speed and coordination to obtain food, mercury may be particularly hazardous to these animals.

Recent research in Minnesota indicates that the following environmental effects are occurring:

- Loons are accumulating so much mercury that it may be affecting their ability to reproduce;
- Elevated levels of mercury have been found in mink and otters;
- Walleye reproduction may be impaired by the fish's exposure to mercury.

Similar effects are being documented for other fish and fish-eating species around the United States and Canada. Has there always been mercury contamination, or is this a recent problem? This is a difficult question to answer, in part because of a lack of adequately preserved fish specimens of preindustrial age to compare against contemporary samples.

However, several lines of evidence from recent studies on Wisconsin lakes suggest that increased emissions to the atmosphere, and subsequent higher deposition rates to lakes, likely translate into higher mercury levels in fish.

The Mercury Cycle and Bioaccumulation

There is a constant biogeochemical cycle of mercury. This cycle includes:

- Release of elemental mercury as a gas from the rocks and waters (degassing); 140 Encyclopaedia of Environmental Science
- Long-range transport of the gases in the atmosphere;
- Wet and dry deposition upon land and surface water;
- Absorption onto sediment particles;
- Bioaccumulation (or biomagnification) in terrestrial and aquatic food chains. Bioaccumulation means an increase in the concentration of a chemical in an organism over time, compared to the chemical's concentration in the environment. Bioaccumulation can be a normal and essential process for the growth of any species, but the accumulation of unnecessary chemicals or toxins, or even the overaccumulation of essential substances can be detrimental. All animals, including humans, daily bioaccumulate many vital nutrients, such as vitamins A, D, and K, trace minerals, essential fats and amino acids, but unfortunately, they can also accumulate many unnecessary substances, such as lead or mercury. What concerns toxicologists is the bioaccumulation of necessary substances to levels in the body that can cause harm.

With substances such as lead or mercury, any accumulation at all can be harmful. Compounds accumulate in living things any time they are taken up and stored faster than they are broken down (metabolized) or excreted. Understanding the dynamic process of bioaccumulation is important in protecting humans and other organisms from the adverse effects of chemical exposure, and it has become a critical consideration in the regulation of chemicals.

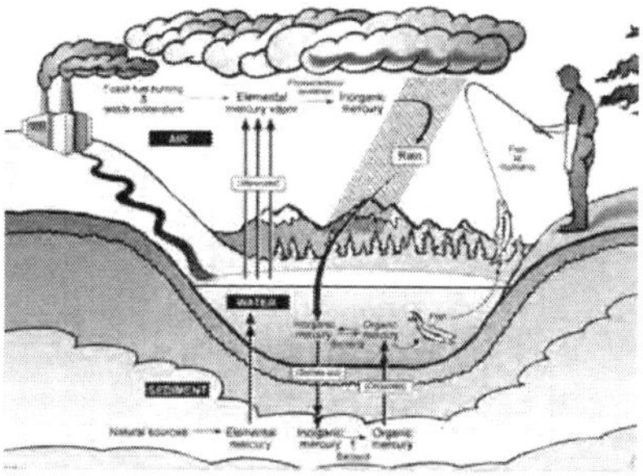

Fig. 4.15: Mercury Cycle

Bioaccumulation varies among individual organisms as well as among species. Large, fat, long-lived individuals or species with low rates of metabolism or excretion of a chemical will bioaccumulate more than small, thin, short-lived organisms. Thus, an old lake trout may bioaccumulate much more than a young bluegill in the same lake.

Above is a schematic drawing of mercury cycling in an aquatic ecosystem. With the exception of isolated cases of known point sources, the source of most mercury to most aquatic ecosystems is deposition from the atmosphere, primarily associated with rainfall.

In the aquatic environment, mercury can be:

- Dissolved or suspended in the water
- Trapped in the sediments
- Ingested by living things (biota)

Methylmercury is the form of mercury most available and most toxic to biota (including zooplankton, insects, fish, and humans). This form of mercury is easily taken up by biota and bioaccumulates in their tissues. Unlike many other fish contaminants, such as PCBs, dioxin, and DDT, mercury does not concentrate in the fat, but in the muscle tissue. Thus, there is no simple way to remove mercury-contaminated portions from fish that is to be eaten.

LIFE AND BIOGEOCHEMICAL CYCLES

The Sulfur Cycle

Another example of a major biogeochemical cycle of significance to climate and life is the sulfur cycle. Living things require certain safe, low levels of this nutrient. The sulfur cycle can be thought of as beginning with the gas sulfur dioxide (SO2) or the particles of sulfate (SO4=) compounds in the air. These compounds either fall out or are rained out of the atmosphere. Plants take up some forms of these compounds and incorporate them into their tissues. Then, as with nitrogen, these organic sulfur compounds are returned to the land or water after the plants die or are consumed by animals. Bacteria are important here as well since they can transform the organic sulfur to hydrogen sulfide gas (H2S). In the oceans, certain phytoplankton can produce a chemical that transforms to SO2 that resides in the atmosphere. These gases can re-enter the atmosphere, water, and soil, and continue the cycle.

In its reduced oxidation state, the nutrient sulfur plays an important part in the structure and function of proteins. In its fully oxidized state, sulfur exists as sulfate and is the major cause of acidity in both natural and polluted rainwater. This link to acidity makes sulfur important to geochemical, atmospheric, and biological processes such as the natural weathering of rocks, acid precipitation, and rates of denitrification. Sulfur is also one of the main elemental cycles most heavily perturbed by human activity. Estimates suggest that emissions of sulfur to the atmosphere from human activity are at least equal or probably larger in magnitude than those from natural processes. Like nitrogen, sulfur can exist in many forms: as gases or sulfuric acid particles. Sulfuric acid particles contribute to the polluting smog that engulfs some industrial centres and cities where many sulfur containing fuels are burned. Such particles floating in air (known as sulfate aerosols) can cause respiratory diseases or cool the climate by reflecting some extra sunlight to space.

The lifetime of most sulfur compounds in the air is relatively short (e.g. days). Superimposed on these fast cycles of sulfur are the extremely slow sedimentary-cycle processes or erosion, sedimentation, and uplift of rocks containing sulfur. In addition, sulfur compounds from volcanoes are intermittently injected into the atmosphere, and a continual stream of these compounds is produced from industrial activities. These compounds mix with water vapour and form sulfuric acid smog. In addition to contributing to acid rain, the sulfuric acid droplets of smog form a haze layer that reflects solar radiation and can cause a cooling of the earth's surface. While many questions remain concerning specifics, the sulfur cycle in general, and acid rain and smog issues in particular are becoming major physical, biological, and social problems.

BIOGEOCHEMICAL CYCLES

The Earth is a closed system for matter, except for small amounts of cosmic debris that enter the Earth's atmosphere. This means that all the elements needed for the structure and chemical processes of life come from the elements that were present in the Earth's crust when it was formed billions of years ago. This matter, the building blocks of life, continually cycle through Earth's systems, the atmosphere, hydrosphere, biosphere, and lithosphere, on time scales that range from a few days to millions of years. These cycles are called biogeochemical cycles, because they include a variety of biological, geological, and chemical processes. Many elements cycle through ecosystems, organisms, air, water, and soil. Many of these are trace elements. Other elements, including carbon, nitrogen, oxygen, hydrogen, sulfur, and phosphorus are critical components of all biological life. Together, oxygen and carbon account for 80 per cent of the weight of human beings. Because these elements are key components of life, they must be available for biological processes.

Carbon, however, is relatively rare in the Earth's crust, and nitrogen, though abundant in the atmosphere, is in a form that is not useable by living organisms.The biogeochemical cycles transport and store these important elements so that they can be used by living organisms. Each cycle takes many different pathways and has various reservoirs, or storage places, where elements may reside for short or long periods of time. Each of the chemical, biological, and geological processes varies in their rates of cycling. Some molecules may cycle very quickly depending on the pathway. Carbon atoms in deep ocean sediments may take hundreds to millions of years to cycle completely through the system. An average water molecule resides in the atmosphere for about ten days, although it may be transported many miles before it falls back to the Earth as rain.

How fast substances cycle depends on its chemical reactivity and whether or not it can be found in a gaseous state. A gaseous phase allows molecules to be transported quickly. Phosphorous has no gaseous phase and is relatively uncreative, so it moves very slowly through its cycle. Phosphorus is stored in large amounts in sediment in the oceans or in the Earth's crust and is recycled back to the surface only over very long periods of time through upwelling of ocean waters or weathering of rocks.

Biogeochemical cycles are subject to disturbance by human activities. Humans accelerate natural biogeochemical cycles when elements are extracted from their reservoirs, or sources, and deposited back into the environment (sinks). Humans have significantly altered the carbon cycle by extracting and combusting billions of tons of hydrocarbons in fossil that were buried deep in the Earth's crust, in addition to clearing vegetation that stores carbon. Global release of carbon through human activities has increased from 1 billion tons per year in 1940 to 6.5 billion tons per year in 2000. About half of this extra carbon is taken up by plants and the oceans, while the other half remains in the atmosphere.

In addition to carbon cycle, humans have altered the nitrogen and phosphorus cycles by adding these elements to croplands as fertilizers, which has contributed to over-fertilization of aquatic ecosystems when excess amounts are carried by runoff into local waterways.

Researchers are trying to understand all of the various pathways and flows of each of the biogeochemical cycles in order to understand how human activities affect these cycles. While many important processes have been understood for more than century, there are many phenomena that scientists are just beginning to investigate. Satellite technology, among other tools, has revealed new information about interactions between the oceans and atmosphere that contribute to knowledge about the carbon cycle, but there remain many unanswered questions.

Human Modifications of Climate Services

Human activities are significantly perturbing all of these biogeochemical cycles as well as other earth system processes, both directly through industrial processes and indirectly through changing distributions and abundance of life. The atmosphere is of particular importance to the perturbations due to its crucial role in mediating all energy that enters and leaves earth. Overall, the atmosphere is the component that controls the dominant energy flow in the earth's climate system, and solar radiation from the sun provides the energy to make the weather machine work.

Embedded in this process are the biogeochemical cycles we have described that operate on a variety of time and space scales and help to regulate flows of energy and materials throughout the earth system Yet, while we understand much about the functioning of separate parts of this system, there is still a great deal to be discovered about the feedbacks and linkages that allow these interconnected parts to function as a whole and, in turn, how they will respond to human modification.

Human Disturbance

Life influences the amount of CO2 in the atmosphere through photosynthesis, respiration, and oceanic absorption. As ecosystems are altered, the balance of these processes will be altered. Human activities are upsetting this balance and increasing CO2 in the atmosphere through the burning of fossil fuels and clearing of forests.A significant increase in CO2 could have dramatic consequences. Mathematical models of the climate suggest that when CO2 doubles (sometime in the middle of the next century should population, economic and technology trends continue as typically projected), the world will warm up somewhere between 1 and 5 degrees Celsius by 2100 A.D. unless other factors counteract or amplify the CO2-induced change Even the lower end of that range is a projected warming at the rate of one degree per hundred years, a factor of ten faster than the one degree per thousand that has been the typical average rate of natural sustained global temperature change from the end of the ice ages to warmer interglacial times. Should the higher end of the 1 to 5 degree Celsius range occur, then we could see rates of climate change some fifty times faster than sustained, natural average conditions. Climate largely determines the types of ecosystems that occupy an area. Global climate change at such a rapid rate would force many species to shift their ranges in an attempt to keep up with changing climatic conditions, as occurred during the ice age-interglacial transition ten to fifteen thousand years ago.

Migrations of species such as slow growing trees with large seeds would have to occur much faster than they did in the past to keep up with rapidly shifting climates. Other species could move more easily, raising the likelihood that communities of species could be disassembled. Estimating the rates of global warming in the next century, however, are very controversial because of the uncertainties involved with multiple interacting feedback mechanisms Humanity can control climate in ways other than changing greenhouse gas concentrations. Consider the amount of moisture released to the atmosphere through transpiration in the tropical rainforests.

The dense vegetation in areas such as the Amazon basin typically recycles the precipitation that falls on it many times over, helping to form heavy cloud cover in the region. The clouds, in turn, reflect sunlight and produce more rain, directly influencing regional climate as well as indirectly perturbing global climate through altering large-scale circulation patterns over the tropics. As humanity deforests regions like the Amazon, not only is CO2 released into the atmosphere, but changes in the hydrologic cycle will almost certainly affect regional climate and possibly even global climatic patterns. In deforested areas of northeastern Brazil, the cutting of the tropical forests has led to Biogeochemical Cycle 145 desertification, changing both surface reflectivity and the rate of transpiration. This change in ecosystem character can lead to a destabilizing positive feedback, which may cause an even further reduction in precipitation. In a recent study on the possible climatic impacts of tropical deforestation, researchers suggest that conversion of forest into crop land or pastures would cause significant changes in the local microclimate Expected changes include reduction in soil moisture, larger diurnal fluctuation of surface temperature and humidity deficit, and increased surface runoff during the rainy season and decreased runoff during the dry season. Results from general circulation model simulations of large-scale deforestation and conversion to grassy vegetation in the Amazon basin indicate an increase in surface temperature, decrease in evapotranspiration, and significant reduction in precipitation.

Depending on the scale of the disturbed areas, local climate changes can lead to regional climate changes which, in turn, may cause alterations in the global climate through atmospheric connections between tropical circulation and large-scale circulation patterns outside of the tropics. The effect on the ecological systems through changes in the hydrological cycle, an increase in the dry season, and the disruption of plant-animal interactions may make it difficult for the rainforests to re-establish themselves if they are destroyed. Climate change aside, the implications of this scenario for the conservation of biodiversity are serious.

The provision of fresh water and regulation of its flows through precipitation, evapouration, transpiration, and run off is mediated by all ecosystems. Forests and other vegetation types are critical components of this ecosystem service providing free flood and drought relief among other things. The loss of these services, through landuse change, can exacerbate disasters like spring floods in the Midwest and Southeast resulting from large expanses of land cleared for agriculture as well as the drainage of wetlands and swamps which otherwise might have acted as reservoirs for holding excess water or filtering toxic wastes.

Valuing Carbon

There is already a historic background on the evaluation of carbon that includes climate change policies such as the introduction of carbon taxes that reduce greenhouse gas emissions through increasing prices of carbon-based fuels proportional to the amount of carbon they emit Another mechanism for reducing greenhouse gas emissions is through an international tradable emissions permit system intended to limit emissions of certain pollutants. These policies and others constitute ways of balancing the economic costs of emissions with somc assumed benefit of averting the loss of ecosystem services (called "climate damage"). Any comprehensive attempt to evaluate the societal value of climate change should include such things as loss of species diversity, loss of coastline from increasing sea level, environmentally displaced persons, and agricultural losses. first estimated the climate damage at 1% reduction in GNP based on market sector losses for a central estimate of climate change. This was criticized as too narrow a view of climate as a type of public good since it reflected neither non-market values (e.g. species loss) nor climate "surprise" scenarios conducted a survey of conventional economists, environmental economists, atmospheric scientists, and ecologists.

Their estimates of loss of gross world product (GWP) resulting from a 3 degree Celsius warming by 2090 varied between a loss of 0 and 21% of GNP with a mean of 1.9%). For a 6 degree Celsius warming scenario, the respondents predicted a loss of the world economy ranging from 0.8 to 62% with a mean estimate of 5.5%. A striking difference was noted between respondents from different academic disciplines, with natural scientists' estimates of economic impact 20 to 30 times higher than conventional economists'. Even a two per cent loss of GWP in 1995, however, represents climate damage of hundreds of billions of dollars annually.

While it is impossible to estimate credibly a numerical value on all of the ecosystem services provided through the maintenance of the carbon cycle at present state, it may be useful to look at land use change and loss of biomass, mostly through deforestation, as a source of atmospheric CO2. In a very simplistic and preliminary evaluation, we can use the rates of net deforestation to calculate a value for carbon. Global loss of above-ground biomass from deforestation in the tropics is approximately 1-3 gigatons/year over the past 10 years.

This amounts to between 2-5 gigatons of carbon in carbon dioxide released into the atmosphere each year from deforestation and forest degradation (this does not include the 6 gigaton carbon emissions from the burning of fossil fuels). Much of the carbon from biosphere emissions is taken up immediately by vegetation, however, leaving approximately 1-2.5 gigatons of net carbon added to the atmosphere each year. We can apply the concept of carbon taxation for emissions to an ecosystem service valuation of retaining the carbon in the forests.

Using a range of carbon taxes from typical macroeconomic models between $1 per ton and $100 per ton of carbon, the net value of carbon lost each year amounts to between $1 and $250 billion/year. However, use of optimizing economic models to estimate climate damage is highly unsatisfying since these studies use very limited and often ad hoc assumptions that both over and underestimate the likely damages to various market and non-market sectors.

Methods of Valuation

The need for alternative methods of evaluation of these climate-related ecosystem services is quite clear when examining preliminary public opinion responses of global warming. In a controversial method called contingent Biogeochemical Cycle 147 valuation respondents are surveyed to determine how much they would be willing to pay to prevent a given global climate change scenario from happening or accept if so much change were to be allowed. The difficulties with this type of valuing of environmental goods and processes are immense, especially since much of the evaluation is subjective. Public opinion depends, in part, upon people's exposure to the issues and the level of education and information on these issues they have received.

In a Southern California study, the contingent valuation technique was applied to the determine the influence of potential changes in temperature and precipitation resulting from global warming on respondents willingness to pay. Factorial survey methods were used to present a variety of hypothetical climate scenarios to a sample of 600 Southern California residents. Respondents were provided with a baseline microclimate for the region before future climate scenarios were evaluated.

For residents living in coastal communities, the baseline climate over the past 10 years was described as having a summer average high temperature of 75 degrees Fahrenheit, with daily highs ranging between 70 and 80 degrees, and an average of 13 inches per year of rain. One possible future scenario over the next 10 years included a summer average high temperature of 100 degrees Fahrenheit, with daily highs ranging from 80 to 120 degrees (the latter typical of Death Valley, California), and an average of 20 inches per year of rain. With these and other scenarios, predicted probabilities were determined from the respondents willingness to pay for the abatement of different mean high temperatures. In this scenario, respondents were willing to pay an average $140 to offset a mean high temperature of 100 degrees, while a mean high temperature of 80 degrees was worth approximately $100 This represents a 40% increment in willingness to pay for a 20 degree rise in temperature and other scenario characteristics. residents reached a plateau in their willingness to pay to prevent 120 degree Fahrenheit mean temperatures as compared to 110 degree Fahrenheit mean temperatures.

However, the actual damages to the L.A. basin residents of mean high temperatures of 110 or more degrees Fahrenheit (which would imply occasional extreme heat waves similar in temperature to Death Valley mean highs) would be orders of magnitude more costly, we believe, than a 100 degree Fahrenheit mean high temperature, as such extreme heat would decimate most existing vegetation and threaten the lives of tens of thousands of elderly and other persons vulnerable to heat stroke.

For just such reasons, Berk and Shulman strongly caution against taking the dollar values from the survey literally or using them in cost-benefit analyses as they confound several source of value including stewardship and altruism. In addition, some of the climate increases were well above the range of current scientific estimates of greenhouse warming.

The survey was not done in conjunction with atmospheric scientists and climatologists who could provide more realistic climate scenarios or ecologists, public health officials, or others who could help the respondents realise what such warming might mean for trees, birds, or people. Contingent valuation of the hypothetical good is possible when people believe the survey scenario. We present this type of evaluation study to highlight how difficult it is to find acceptable methods to place values on the climatic components of ecosystem services. In this survey case, the background of the respondents as well as their (limited) prior knowledge of the impacts of greenhouse warming played a large role in the survey outcomes.

At the same time, however, contingent valuation points out that people are willing to pay to preserve ecosystem services as well as the tremendous need for education to help citizens more realistically value climate and other environmental

The ongoing disturbances of the atmosphere that affect the biogeochemical and physical processes that determine the climate may influence human and natural systems in profound ways. We have attempted to outline a few of the major ecosystem services that are associated with climate and the atmosphere, as well as introduce the challenging task of quantifying, and ultimately monetizing, these services.

Current monetized estimates of climate damage by the middle of the 21st century from typical climate change scenarios range from slight economic benefit to a trillion or more dollars lost annually, with most macroeconomic assessments assuming a one to two per cent annual loss to GWP from climate change. Moreover, the interacting processes and biogeochemical cycles occurring across a wide spectrum of scales lead to synergistic effects that are not usually considered and sometimes not even known when we attempt to disaggregate and value ecosystem services.

The deforestation of the Amazon basin is one example of interacting scales where land use change affecting local and regional climate may also produce a net global residual. Even if the mosaic of regional effects average themselves out globally, there could be residual effects arising from heterogeneous forcing of the climate in areas outside of the tropics (i.e. regional high concentrations of sulfate aerosols or tropospheric ozone).

Ecosystems both mediate and respond to the climate system through a variety of physical, biological, and chemical feedback cycles. The uncertainty of resulting synergisms and potential global effects, as exemplified in the Amazon basin, points to the important challenge of defining and understanding the processes that link species and ecosystems with climate. With increasing knowledge, we can better anticipate ecological responses under changing climate scenarios. Meanwhile, humanity continues to perform this potentially trillion dollar unnatural experiment on "Labouratory Earth".

Air, Climate and Weather

The Earth's atmosphere is a blanket of gases approximately 350km (218 miles) thick. It is a large and complex system that interacts with the Sun, the land, and the oceans in order to produce both the Earth's weather and climate.

The Earth's atmosphere has four distinct layers

- Thermosphere
- Mesosphere
- Stratosphere
- Troposphere

Thermosphere

The layer of atmosphere most distant from the Earth is the thermosphere, which begins approximately 80km in altitude. It is also the hottest layer; "thermo" being Greek for heat. The temperatures in the thermosphere increase with altitude due to the absorption of intense solar radiation by the limited amount of remaining molecular oxygen. The source of this heat is through bombardment of solar particles carried on the solar wind that do not reach deeper into the atmosphere.

Mesosphere

The mesosphere extends from 50 to 80km in altitude with very sparse atmosphere, accounting for only about 0.1 per cent of the mass of the atmosphere as a whole. Temperatures decline within the mesosphere as altitude rises, containing the coldest temperatures within the Earth's atmosphere. At its upper boundary, the mesopause, average temperatures are near -110°C in the summer and -60°C in the winter.

Stratosphere

The stratosphere extends from approximately 10-12 km to around 50 km above the Earth's surface. The air temperature remains relatively constant up to an altitude of 25 km, then increases gradually having a stabilizing effect on atmospheric conditions. The stratosphere contains nearly 90 per cent of the atmospheric ozone, which plays a major role in regulating temperatures as solar energy is converted to kinetic energy when the ozone molecules absorb ultraviolet radiation, resulting in the heating of the stratosphere.

Troposphere

The troposphere is the layer closest to the Earth's surface, containing more than 80 per cent of total atmospheric mass composed of 78 per cent nitrogen, 21 per cent oxygen, other trace gases, water droplets, dust, and other particles. The troposphere is where most weather occurs; the circulation of air in intensive vertical movements results in the formation of clouds, while horizontal movements results in wind. Both temperature and water vapour content decrease rapidly as altitude increases within the troposphere. Nearly 99 per cent of atmospheric water vapour is contained within this layer, which plays a major role in regulating air temperature as it absorbs solar energy and thermal radiation from the planet's surface.

Scientists believe that the Earth's climate is changing and is, in fact, heating up. However, there are considerable differences among the views with regard to the rate of change, the impact on our environment, and what can or should be done about it. Current environmental concerns are increa-singly focusing on the particles and gases released into the atmosphere as a result of human activity. Frequently, unusual weather events, such as Hurricane Katrina in 2005, are being linked to global warming. But "weather" and "climate" are two very different things, and it is important to understand the difference, as well as the relation, between the two.

In the troposphere, air rises as it is heated by the sun, falls earthward as it cools, and intermixes with evapourated water from the planet's bodies of water to form clouds and precipitation. The uneven heating of the Earth's surface by the sun, along with Earth's rotation, creates rising (convection), falling (advection), and horizontal air movements (winds). The result of these processes occurring in the form of rain, snow, heat or freezing cold, at a particular place and time, is called weather. The longer term trends in patterns of temperature, rainfall, and other weather indicators over time, usually in blocks of 30 years that can affect the entire Earth, is called climate.

Land and Land Use

Land is the solid part of the Earth's surface, also known as the lithosphere. The surface of the Earth is shaped by a combination of physical processes, including earthquakes and volcanoes, shifts of rocks and sediments, and flows of river and ice. Human activities also shape the land in many ways, including expanding agriculture and population, excavating mineral and forest resources, and changing the flow of rivers with dams and channels.

Land cover is the physical and biological material found on the land surface – which can be either vegetation or man-made structures. Land cover differs from land use, although the way land is used is often either a cause or an effect of its cover type. Land use describes the various ways in which human beings make use of and manage the land and its resources. The way land is used and managed can alter land cover, and can also have an impact on surrounding habitats, the flow of surface waters, and on biogeochemical cycles.

Land cover and land use issues, therefore, can play a significant role in environmental change, including climate change. When land cover is altered, it can result in the destruction of natural carbon sinks that absorb atmospheric carbon dioxide, a heat-trapping gas that contributes to climate change. Deforestation, is a major contributor to observed increases of carbon dioxide in the atmosphere.

There are often competing interests and tradeoffs related to land use that must be taken into consideration. Land use planners and decision makers often struggle with how to meet human needs for food and the utilization of our natural resources while also minimizing impacts and preserving important habitats. These choices are typically made by those who own or control the land; although the choices can also be limited by the physical and biological characteristics of the land, including the climate, soils, and geography.

Land use choices are also limited by institutional factors and, in the United States, can be governed by a mix of local, county, state, and federal laws. Although most of the land. is privately owned, the federal government is the single largest owner of land. Analysis of land use and land cover has been greatly facilitated in the last few decades with the availability of new technologies, including remote sensing. The development of geographic information systems (GIS) has made it possible to combine physical characteristics of the land with information about population, economic data, and other information for further analysis

Water

"Water, water, everywhere, nor any drop to drink." More than 70 per cent of the Earth's surface is covered by water. When the Apollo space missions beamed back pictures of the Earth, public attention was caught by the image of a "Blue Planet," because of the vast expanses of ocean covering the globe. Yet despite this seemingly inexhaustible supply (approximately 370 billion billion gallons) of water on earth, access to clean drinking water remains a major challenge for a large percentage of the world's population. Almost 97 per cent of the Earth's water is in the oceans and is too salty to drink. Less than 1 per cent of the water on the earth is fresh water (the remaining two to three per cent is frozen fresh water contained in glaciers and ice caps).

Water is indispensable for life, not only for drinking water, but also for raising crops and animals for food. Fishing provides a major source of protein for much of the world's population. Throughout history, cities and villages have grown up near sources of water, for drinking, sustenance, and transportation. In this century, population growth has increased in areas where fresh water is relatively scarce. Water in these areas is a valuable commodity and, in some cases, great technological feats have been undertaken to provide a reliable source of water. In some areas, water quality is a serious concern and waterborne diseases remain a significant source of mortality in many developing countries. The United States has made considerable progress over the last few decades in reducing direct discharges of pollutants into lakes, rivers, and streams. However, indirect sources of pollutants, including both nutrients and toxic substances that are carried by flowing surface waters into larger bodies of water, have proven more difficult to control.

Ecosystems

An ecosystem is generally defined as a community of organisms living in a particular environment and the physical elements in that environment with which they interact. But where does one particular ecosystem end and another begin? An ecosystem can be as small as a field or as large as the ocean, depending on the scale that the researcher is examining. The borders of an ecosystem may be clear, such as a pond; other borders may be less easy to define, such as grassland that gradually changes into brush.

Just as there is an immense diversity of individual species on the planet, so is there a rich diversity of ecosystems, from the icy arctic zones to tropical forests lush with plants and animals. Even the depths of the oceans, once thought to be barren, are now known to be teeming with living microorganisms and other life. There is much that remains to be discovered. Biologists do not know with any certainty how many species there are or even why some areas, such as the tropics, are richer in biological diversity than others.

It is known that human activities, mostly unintended, threaten biodiversity by altering habitats and introducing non-native species. There is general agreement on the importance of protecting biological diversity, especially since humans depend on the services provided by living organisms and ecosystems. There is less agreement on the best approaches to conservation and how to balance preservation of habitats with meeting human needs.

Energy

The development of modern civilization has been dependent on both the availability and the advancement of energy. We have witnessed a progression from animal and steam power, to the internal combustion engine and electricity generation and to the harnessing of alternative sources of energy. Because of our reliance on energy sources, it is also important to understand the impacts of energy use on the environment. All aspects of energy the way it is produced, distributed, and consumed can affect local, regional, and global environments through land use and degradation, air pollution, the acidification of water and soils, and through global climate change via greenhouse gas emissions.

The majority of our current energy stems from fossil fuels such as coal, oil, and natural gas. Coal dominates in our production of electricity, while oil is the world's primary transportation fuel. Natural gas use, which is most commonly as heating fuel, is currently growing faster than other fossil fuels. While cleaner and less carbon intensive than both coal and oil, its use still emits significant amounts of carbon dioxide.

Over the foreseeable future, it is very likely that fossil fuels will remain our largest source of energy. However, fossil fuels are finite resources and there is concern not only about both domestic supply and reliance on foreign supplies, but also with the increasing cost of these fuels. The burning of fossil fuels is also the largest source of carbon dioxide emissions, which contributes to the greenhouse effect. As these fuels continue to decrease in supply and impact the environment, there will be an increasing need to utilize alternative fuel sources.

Nuclear energy provides nearly a fifth of the world's electricity and does not produce harmful by-products, only non-radioactive water vapour. However, controversy remains over the use of nuclear energy due to concerns over safe storage and disposal of radioactive waste, and the potential for accidents and radiation contamination and exposure.

Hydrogen is the most abundant element in the universe and could be an important factor in our energy future. Hydrogen fuel cells can produce power without emitting any pollutants; their only byproducts are water and heat. Hydrogen can both carry and store energy and can be used in a wide variety of applications, including portable devices that use batteries, transportation vehicles, and a number of stationary power sources. The majority of hydrogen is currently produced by processing fossil fuels (which emits pollutants through the initial process), but it can also be manufactured from renewable energy sources and feedstocks through the electrolysis of water.

With the increased cost and environmental impacts related to fossil fuel use, along with the controversy over nuclear power, research and development in the area of renewable sources of energy continues to flourish. Alternative sources – such as wind, solar, geothermal, and water – have been used in one form or another for many centuries, but will require additional advancement before they can become truly cost-competitive with conventional energy sources.

Wind is an abundant, inexhaustible, and clean source of energy; and, advancements in the design of wind turbines has lowered the cost of windgenerated electricity. However, wind variability, noise pollution caused by the turbines, and the potential to harm birds continue to be areas of concern.

Solar technology obtains power from sunlight. There are three types of solar technology: solar electric, solar hot water, and concentrating solar power. The cost of solar power has declined over the past 30 years, and will continue to decline with increased technological efficiency of technology and manufacturing improvements. However, it is also subject to changes in weather (cloud cover) and pollution, which can block sunlight and decrease overall productivity.

Geothermal energy taps into heat underneath the Earth's crust to boil water that is then used to drive electric turbines to heat buildings, homes, or in other non-electrical purposes. Over the past two decades, geothermal energy costs have decreased nearly 50%. However, while abundant, in only a few locations is it close enough to the Earth's surface to provide for large scale power.

Water–or hydroelectric–power has been harnessed for thousands of years, beginning with the water wheel. For more than a century, the technology for using falling water to create hydroelectricity has evolved to its current use of turbines and generators to convert the energy into electricity, which is then fed into the electric grid. Although a widely accessible energy source in many countries–concerns over land use, impacts on fish populations, and water flow and quality remain.

One effective way of reducing environmental impacts from the current production of energy is to increase energy conservation. By using energy more efficiently, both costs and emissions are lowered; therefore, a significant incentive exists to improve efficiency in manufacturing processes, the construction of new buildings, the production of new appliances, and for many other energy uses.

Several developed countries are already taking action to increase energy efficiency and, on a larger scale, new measures are being explored to assist developing countries with energy conservation as they continue to increase their energy consumption. Under the Kyoto Protocol, developed countries can provide technological assistance through "clean technology" projects with developing nations and can even put the emission reductions toward compliance in their own countries.

As human consumption of energy increases and we come closer to using up our supply of fossil fuels, further research and development will be necessary to produce alternative and/or renewable sources of energy that are readily available, affordable, and less harmful to the environment. While our dependence on energy is not likely to decrease, there will continue to be an increase of new innovations in energy technologies with a larger focus on energy conservation and efficiency.

Energy Information Administration (EIA): The EIA also provides information on the various types of energy, including renewable, non-renewable, and secondary sources. Congressional Research Service: Energy Issues Policy papers on energy issues can be found on this site, including articles and several reports related to energy policy and environmental protection.

Food

Human beings began domesticating plants and animals as long as 10,000 years ago. Success in this endeavor led to drastic changes in how and where human beings lived. It changed how humans interacted with each other and how they interacted with the earth. Agriculture led humans from a nomadic existence to one based in permanent and semi-permanent settlements. Human civilization has its roots in the early domestication of plants and animals.

The unprecedented growth in global population that occurred in the twentieth century was made possible by remarkable advances in agriculture and technology. The Green Revolution brought high-yield crops and advanced growing techniques to developing countries and improved nutrition and health for most parts of the world. Continuing population growth will require more acres of land to plant and more water for irrigation, increasing pressure on habitats and natural resources.

Advanced agricultural methods have made it possible to grow more food on fewer acres of land, which permits much land to be returned to forest and other uses. However, there are costs associated with high-yield methods, including their heavy reliance on pesticides and fertilizers. Runoff of fertilizers from agricultural lands has affected the water quality of some lakes, rivers, and bays.

Erosion is, and has been since crops were first planted, a serious environmental problem in the absence of preventative measures. Irrigation, required in many parts of the world, can be a significant drain on water supplies in arid areas and, if improperly managed, can lead to buildup of salt deposits, which may severely degrade soil quality.

One of the most critical challenges of the next few decades is to find ways to increase production of food while minimizing environmental degradation. Considerable research in this and other countries has led to advanced methods to preserve soil and prevent erosion, including conservation tilling and computer-controlled application of pesticides, fertilizers, and water. Pesticide use is controversial. Other methods, especially new types of genetic modification, or "genetic engineering" are more controversial. To reduce reliance on pesticides, methods such as biocontrol (using other species to control pests) and biotechnology to create pest-resistant crops are being tested. Biocontrol using wild-type (naturally occurring) species is usually accepted, especially by organic producers. Naturally occurring mutants are accepted as well. More controversial is the use of genetically modified biocontrol agents changed by DNA deletion or insertion of new traits. Concerns about the potential impacts of these methods have been raised.

Environment and Society

Throughout history humans have both affected, and been affected by, the natural world. While a good deal has been lost due to human actions, much of what is valued about the environment has been preserved and protected through human action. While many uncertainties remain, there is a realization that environmental problems are becoming more and more complex, especially as issues arise on a more global level, such as that of atmospheric pollution or global warming. Interactions between human society and the environment are constantly changing. The environment, while highly valued by most, is used and altered by a wide variety of people with many different interests and values. Difficulties remain on how best to ensure the protection of our environment and natural resources. There will always be tradeoffs and, many times, unanticipated or unintended consequences. However, a well-managed environment can provide goods and services that are both essential for our well being as well as for continued economic prosperity.

❐

5

Fogs and Clouds

Such ordinary aerological processes as the formation of rain are now understood almost as well as any laboratory experiment. First causes in the sun or elsewhere may still be obscure, but the general earthly mechanism of fogs and clouds and showers is reasonably clear.

MOISTURE VISIBLE AND INVISIBLE

Invisible water vapour is always present in the air, varying in amount from a mere trace up to about four per cent by volume or two and one half per cent by weight (the water vapour being somewhat lighter than dry air). 'Specific humidity'—the actual mass of water vapour present in a given mass of air, measured usually in grams of water vapour per kilogram of air, and varying from less than 1 g/kg in very dry air up to perhaps 20 g/kg in very moist air—is the most conservative and generally useful measure of atmospheric moistures.

For some purposes, though, dryness or moistness of air is more conveniently expressed as 'relative humidity'—the ratio of the actual amount of water vapour present (the existing specific humidity) to the greatest possible amount of water vapour that could exist at the existing temperature (the saturation specific humidity).

And for many practical weather purposes, the best measure of atmospheric moisture is the 'dewpoint'-that temperature which has a saturation specific humidity equal to the present, existing specific humidity. Finally, we saw how any one of three conditions indicates that condensation of water vapour into liquid form is imminent:

- The nearness of the existing specific humidity to the saturation specific humidity;
- The nearness of the dewpoint to the temperature;
- The nearness of the relative humidity to one hundred per cent.

When water vapour in the air comes out of its invisible form into the visible form of liquid water, however finely divided, it must have some solid nucleus on which to form. Now, on a summer night, say, if the lower air is quite still as the dark earth radiates away its heat, the cooling ground cools only that layer of air immediately next to it.

When the temperature of this still, surface air reaches its dewpoint, invisible moisture condenses into liquid dew on sticks and stones, blades of grass, or any other cool and convenient objects—nuclei of condensation, we might call them, albeit large ones. Thus the fairly common idea that 'dew falls' is wrong. If the temperature happens to be below freezing (less than +32°F or 0°C), the invisible water vapour sublimes directly into feathery ice crystals, forming hoar-frost.

Invisible water vapour that condenses into minute droplets in the free air, forming fog or clouds, also needs solid nuclei on which to form. But whereas the nuclei for dew and frost are large and evident, nuclei for fog and cloud droplets are of infinitesimal size. Nevertheless these nuclei—minute particles of sea salt, smoke, hygroscopic dust, and the like—are always present in the troposphere. Given these nuclei, rapid condensation of invisible moisture into minute droplets of liquid water (visible in the aggregate as fog or cloud) begins when, by reason of evaporation, dewpoint increases to existing temperature; or when (as is far more usual), by reason of atmospheric cooling, temperature reaches existing dewpoint. In any case, condensation begins when and where dewpoint and temperature come together, for any reason, at any level in the air ocean.

FOG

Fog can be at best a creation of elfin beauty, is on the average an inconvenience scarcely ever welcomed by honest people, and becomes at worst, in transportation, a great hazard. It may stall land vehicles for hours, force ships along the coast to anchor for days, and even more sorely bedevil airplanes that must keep moving at two or three miles per minute if they are to keep the air at all. In temperate latitudes the gray, creeping fog-mass is usually formed entirely of myriads of minute water droplets, each something like one thousandth of an inch in diameter.

Damp haze, which in its denser forms may approach the murkiness of light fog, can (unlike fog) begin forming when temperature and dewpoint are far apart. This denser form of haze sometimes extends for several thousand feet above the earth's surface, and the well-defined top of the haze layer usually marks the base of a temperature inversion. To an aviator above the haze it appears almost as a level, pearly-white sea brilliantly reflecting the sunlight; only as he descends into the haze do ground patterns begin to loom murkily into view. But at night, lights shine upward clearly through this haze, whereas true fog of any considerable thickness blots them entirely from view. The reason for this is that fog droplets are much larger than damp-haze droplets (though still incredibly small)—large enough, in relation to the wavelength of visible light, to block off almost completely any direct rays. Yet the transition from dense haze to light fog may be gradual, with gradually increasing relative humidity. As dense fog requires both plenty of vaporous moisture to start with, and then considerable cooling, for its formation, it is found mostly in temperate latitudes. In arctic regions, or at high altitudes, fog (or cloud) may be composed of myriads of minute ice-needles—'frost smoke' that glitters and sparkles in the sun. In the tropics fog occurs but rarely, because the temperature contrasts that favour it are usually lacking.

Fog is classified, according to the mode of its formation, into three main types: (1) radiation fog, also called 'land fog' or 'summer fog,' in which air already stagnant on the scene cools until its dewpoint is reached; (2) advection fog or transportation fog, also called 'winter fog' or 'sea fog,' in which air is brought in bodily from outside regions to the scene of its fogginess; and (3) frontal fogs, usually temporary and transient, that occur with the passage of fronts between unlike air masses.

Radiation Fog

Radiation fog is most frequent during the long, clear nights of autumn, though it may occur at any season, particularly in rather stagnant air conditions near the centre of a high-pressure area. The fog sometimes appears to rise from the ground; but actually, the only thing that rises is a zone of temperatures lower than the dewpoint of the air in question. As the ground and the moist air immediately above the ground slowly radiate their warmth away into space, the temperature of this lower air falls, approaching its dewpoint. When the two coincide, fog appears in visible form. As the top of the fog layer itself radiates away heat much faster than dry air, the fog usually continues to deepen until sunrise.

On a flat plain, if the wind is very light (say only one or two miles per hour), the fog layer may be very shallow. This 'ground fog,' perhaps only a few feet thick, results from very gentle mixing in the air immediately above the ground; there is not enough eddying and turbulence to raise the surface coldness higher. But in rolling or mountainous country, even on a calm night, cooled surface air drains into the hollows (as anyone can notice by driving over darkened hill and dale in an automobile, or even walking across broken terrain, with a thermometer). As a result, large pools of valley air may be cooled sufficiently to produce fog 'lakes' many fathoms deep.

If the breeze is somewhat stronger—say two to six miles per hour—the surface coldness is mixed by eddying and turbulence through a considerable layer of the lower air, causing true fog that blots out hillocks and hollows alike. This sort of fog usually begins at or near the surface, working gradually upward, though to an observer on the ground it seems to materialize rather suddenly out of nothing more tangible than a few stray wisps of vapour. The maximum height of the fog depends on many aerological factors, but it is seldom more than a few hundred feet. Temperature in the fog layer usually decreases with altitude at something like the moist adiabatic rate (-2½°/1000 ft); and just above the top of the fog there is usually, if not always, a temperature inversion.

If the breeze is much stronger than six miles an hour or so, eddying and turbulence diffuse the surface coldness through such an extensive layer of overlying air, and bring down so much potential warmth from aloft, that the air temperature never reaches its dewpoint and fog is entirely prevented. The conditions favourable to deep and bothersome radiation fog are, therefore: to begin with a fairly high dewpoint for the season (plenty of actual water present in vaporous form, that is), and not much difference (say less than 15°F at sunset) between temperature and dewpoint (or a corresponding small difference, say less than 7°F, between wet- and dry-bulb temperatures); second, a clear or nearly clear sky, for maximum radiational cooling; and third, through at least part of the night, light breezes registering two to six miles per hour at the surface.

Once begun, radiation fog continues to deepen through the hours of darkness and, curiously enough, may increase for a brief time after sunrise. This last effect may be due to either of two factors: the power of sunlight to make floating particles attract water; or the increase of lower-level turbulence and mixing caused by solar heating at the surface. It is a rather frightening experience to take off from a clear airport at dawn, and see it fog in under you a few minutes later.

Within a few hours after sunrise, however, an ordinary radiation fog begins to 'burn off.' This burning off always proceeds from the ground upward as the solar radiation, penetrating through the fog layer, warms first the ground itself and then, progressively, the air above it. The lower fog does not actually 'lift' in this case any more than it 'rises' out of the ground at night; what actually lifts at first under the morning sun is a zone of temperatures higher than the dewpoint. But before long rising currents from the warmed ground break through the fog layer here and there as the solar radiation continues. And the sun, shining through these breaks with redoubled energy, soon produces enough warm updrafts to raise the whole fog layer slowly upward and sweep it into scattered drifts of cloud.

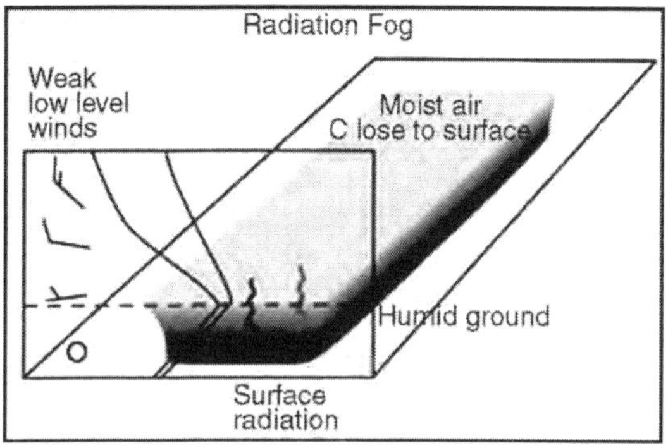

Fig. 3.1: Radiation Fogs

The worst radiation fogs are found around great cities, where innumerable smoke, oil, and soot particles impair visibility on their own account, serve as ready nuclei for condensation, and hinder the evaporation of fog once formed. Thus the infamous London fogs, which greatly increased during the late nineties with the growth of the great city, may hang on in yellowish 'peasoup' form, perhaps for days, until the advent of a new air mass and freshening winds.

Though less tenacious, radiation fogs are nearly as common in many regions of the open country. They can also occur over limited areas of calm ocean, but the usual sea fog arises out of different causes. An extreme example of radiation fog occurs in the Great Valley of California, where murky-gray obscurity may hang on for days beneath the 'lid' of a temperature inversion.

A particular type of radiation fog, called 'high fog' or 'stratus fog,' starts in mid-air and builds downward. This high fog, also, is particularly prevalent in the walled-in valleys of California, but I have noticed it occasionally over the eastern seaboard, and it probably occurs widely, though rather rarely, in other regions.

Once we had a formation of army airplanes operating all night from an air base in the District of Columbia. This high fog formed suddenly, around midnight, at a height of about fifteen hundred feet, trapping one of the planes above it. The predisposing aerological condition here is a zone of decreasing temperature and high humidity from the ground upward for the first thousand feet or so, with a pronounced and lid-like temperature inversion above. The air below the inversion is thoroughly mixed, equally moist, and coldest at the base of the inversion in mid-air. In other words, specific humidity is fairly constant (dewpoint decreasing about 1°F/1000 ft) throughout the moist lower layer, but temperature is lowest at the top.

As the whole moist layer cools by radiation at night, the top temperature reaches its dewpoint first, and the fog builds downward from the base of the inversion. Sometimes you can see it beginning in wisps along the side of a steep mountain. Its downward advance is usually rather sudden at first— perhaps five hundred feet or so in an hour; then it lowers more slowly. But within a few hours, it may reach down to the ground. Some typical temperatures, in a California fog of this sort, are as follows: 600 feet (surface), 57°F; 900 feet (top of fog), 54°; 1200 feet, 64°; 1500 feet, 66°.

Advection Fog

Whereas radiation fog is due to simple cooling of the air in place, advection fog may be due to either of two causes. One is the horizontal advance of warm, moist air over a cold land or water surface. The other is the advance of cold air over a moist, warm surface. Of these two, the first cause is by long odds the most common.

On land, a warm-air advection fog often results when gentle, moist southerly winds blow in from the ocean, particularly if the land surface is covered with snow. The warm, moist air is chilled by the cold ground below it until its falling tempera- ture, uniformly distributed through several hundred feet of depth by gentle turbulence, reaches its dewpoint. In the United States any large northward outbreak of warm, moist (tropical maritime) air in wintertime may produce widespread advection fog.

The warm sector of a marked temperate cyclone may be completely blanketed by such a fog, and if the pressure gradients are steep, this may be one of the rather exceptional cases where fog accompanies a high wind, usually from some southerly quarter. These warm-air advection fogs are far less common on land than radiation fogs; but the advection fogs are likely to be more widespread and prolonged, forming or advancing at any hour of the day or night.

On the ocean, warm-air advection fogs are by far the most common type, and more frequent in summer than in winter. They are likely to form wherever a cold ocean current or area is surrounded by considerably warmer water— such as off Newfoundland, where the Labrador current projects southward; off Nantucket, where the colder water from the northern New England coast meets the warmer waters from south of Cape Cod; or along the whole line of the Pacific coast, where water upwelling inshore from chill depths below is much colder than the Pacific waters offshore.

The cold-water and fog area may be small and distinct—such as the fivemile- wide cold current in the middle of Santa Barbara Channel. The whole fog area moves with the prevailing winds, which are usually light -say fifteen miles per hour or less on the open water—though in a great temperate cyclone they may be much higher.

I remember one case of moderately thick fog on a steamship off the Atlantic coast, with the wind blowing a whole gale out of the southeast and a mountainous sea running. If the wind is much above twenty miles per hour, however, the minute fog droplets tend to coalesce into the larger droplets of mist or drizzle, with somewhat improved visibility.

These sea fogs vary widely in height—sometimes they are so shallow that the mastheads of ships project into clear air above. At other times they are more than a thousand feet deep. Their average depth is perhaps around five hundred feet. In any actual case, of course, this height can be determined by a kite or captive-balloon sounding from a ship in the fog—or from shore elevations along a mountainous coast like California. Sometimes the sea murk may take the form of 'high fog' or low stratus, with good visibility along the water surface.

The second type of advection fog occurs when cold air flows over a moist, relatively warm surface such as ocean or lake water, or possibly land wet by recent rain or melting snow. Here the invisible water vapour, evaporating from the water surface, is chilled below its dewpoint by the cold air above so that it visibly 'steams.' On land, the steaming of lakes on autumn mornings, as radiation-cooled land air sweeps across water areas less prodigal with their summertime heat, shows the mechanism of cold-air advection fogs in miniature form. Similar steaming fogs occur in arctic regions when cold, belowzero polar air sweeps across leads of open water in the pack ice; though here, of course, the fog is formed of ice crystals.

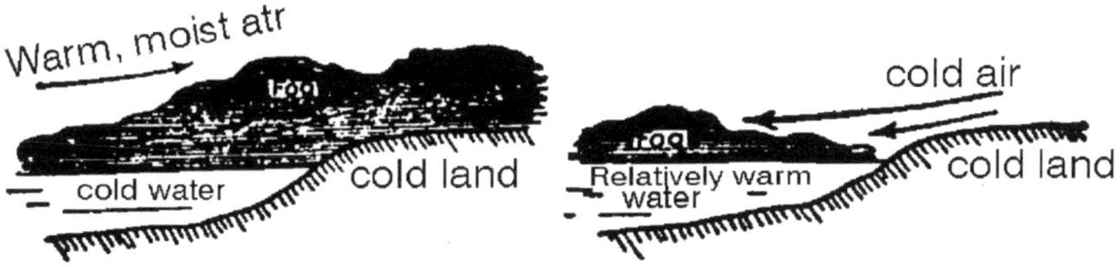

Fig. 3.2: Advection Fogs

Cold-air advection fog on the ocean is quite rare, but notably hazardous when it does occur. Near the United States it is confined mostly to the cast coast, and arises from the rapid southward and eastward outbreak of a large, cold (polarcontinental) air mass. When the fresh polar air, at near-zero temperature, rushes southward or southeastward behind a marked cold front over water at temperatures around 50°F (or even 70° in the Gulf Stream), steaming fog arises on a widespread scale and perhaps masthead high. What with the viewless murk, the piercing cold, and a howling gale out of the north or northwest, even a steel steamship may do well to anchor inshore, despite every modern aid to navigation.

Frontal Fogs

Somewhat different in mechanism from radiation fogs and advection fogs are the fogs that occur near the marked 'fronts' between unlike air masses. These frontal and pre-frontal fogs, however, are neither very common nor very persistent, and only two varieties of them are worth mentioning here.

First there is a type of pre-frontal fog that occurs in advance of a warm front, in a zone perhaps fifty or a hundred miles wide where heavy rain from the warm-front sloping cloud-surface above has saturated the cooler air of the underlying cold wedge, and falling pressure ahead of the front causes further expansional cooling of the cold, moist air. (This pre-frontal fog is distinct, of 5 course, from the warm air mass [advection-type] fog sometimes present in the warm sector behind the warm front.) The other type of pre-frontal fog may occur a few miles ahead of an advancing cold front, where a rapid fall in pressure encourages the fog-forming proclivities of the warm air mass.

Any fog, however formed, may drift bodily elsewhere and engulf regions well removed from its origin, provided aerological conditions do not favour its rapid evaporation. The front of the fog, particularly at sea, may take the well-defined form of a steeply sloping 'bank,' with a distinct and clear-cut edge. When a sea fog comes ashore, it remains on the ground if the land is colder than the water. If the land is relatively warmer, however, the lower surface of the fog lifts somewhat above the ground.

FORMATION OF CLOUDS

Fogs and clouds are of the same texture, and their different appearances are mainly a matter of viewpoint. High cloudturrets may gleam as white, as apparently solid, as alabaster in the bright upper sunlight; but explore the insides of those seeming-solid vapour-peaks in an airplane, and they become just as ragged, gray-murky, cold, and damp as any fog. Having seen how cooling of humid air near the ground produces fog, we may well guess that cooling of some sort in the upper air produces cloud.

How is this large-scale free-air cooling brought about? Not by conduction (for air is a very poor conductor of heat), not usually by radiation (for dry air neither absorbs nor radiates heat readily), but mainly by the expansion cooling of air forced in some way to rise. In the preceding we examined the underlying mechanism of this rising-refrigeration process we followed the salient facts of dry-adiabatic cooling, of moist-adiabatic cooling, of dry-air and moist-air stability and instability. It remains now to see how these various aerological conditions play their part in forming clouds out of thin air.

Forced Cloud Formation

If an entire layer of moist air is driven by prevailing winds up the slope of a mountain range, or drifts up the inclined plane of an underlying wedge of colder air, it cools throughout its whole extent at something like the dry adiabatic rate (about -5°F/1000 ft—or 5.5°, to be exact) so long as moisture remains entirely and invisibly vaporous. But at some moderate altitude, if its humidity is high, the rising air layer reaches its dewpoint. Then moisture begins to condense on myriads of infinitesimal nuclei that are usually present in air— salt particles, haze particles, and so on—into the form of innumerable small water droplets visible in the aggregate as cloud. The individual droplets forming the cloud are about the same size as fog droplets or somewhat smaller—say 1/2500 of an inch in diameter. After this cloud condensation begins, of course, released latent heat retards the further cooling of the rising air to the moist adiabatic rate (say -3°F/1000 ft at moderate altitudes).

If the moist, rising air layer is ill-disposed towards updrafts and downdrafts within itself—if, in a word, it is stable—it forms merely one great, shapeless horizontal mass of flat, quiescent, layer-like cloud in the same way that a groundcooled air layer forms fog. A large proportion of all clouds are formed, as simply as fog, within such moist but stable air layers lifted in some way by external causes. If all clouds were so formed, the science of aerology would be much simplified. But more complicated mechanisms of cloud formation produce some of the most splendid clouds and spectacular storms.

Free Cloud Formation

It is worth recalling from the stability in the preceding that, in 'absolutely unstable' air (lapse rate steeper than about -5°F/1000 ft), vertical currents tend to continue and to grow, once they are triggered off by commonplace causes such as surface heating, mountain updrafts, and so on. The same may be true of 'conditionally unstable' air (lapse rate between -5° and -21/2°F/1000 ft), which behaves like absolutely unstable air once condensation has begun. By the 'beginning of condensation,' of course, we mean the formation of the minute water droplets that constitute visible cloud. The ultimate result of instability in any air layer, dry or moist, is updrafts (together with compensating downdrafts elsewhere). And the result of updrafts, in any moist air, is bound to be clouds.

Unlike the flat, quiescent layer-like clouds formed by general air lifting under external forces, these cloud-children of instability are usually lumpy, heaped-up, puffy, and cauliflower-like in appearance. The instability clouds themselves may be very large, occasionally towering higher than the highest mountains, yet they tend towards localized lumps, turrets, and masses, called 'cumulus,' rather than widespread and shapeless layers, called 'stratus.'

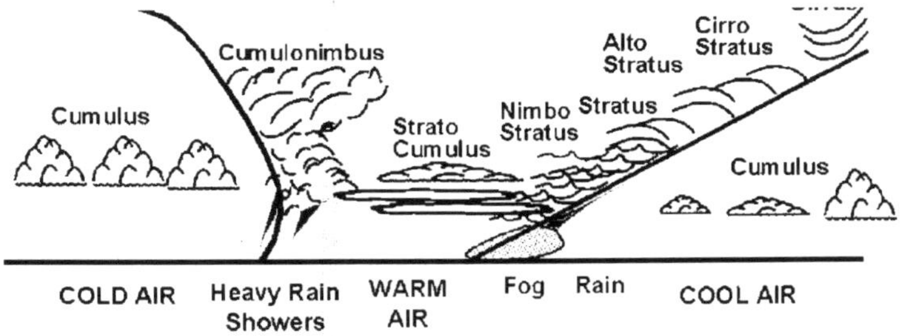

Fig. 3.3: Cloud Formation

In freely rising air, instability cloud-birth always begins at a certain definite level—the condensation level where falling temperature meets existing dewpoint. Under some conditions this condensation level is very definite, and readily apparent as a flat cloud base or 'ceiling.' The typical woolly, cream puff clouds of fair-weather summer skies, for example, all have flat bases (and not the rounded bases that some artists like to paint into them); and these bases are actually all at the same elevation, though more distant clouds may appear lower to ground observ- ers by reason of perspective. Each detached, puffy cloud in this idealized summer picture represents, of course, the visible crown of an updraft (invisible below the condensation level) that is sustaining the cloud. In the clear spaces between these detached woolpacks there are usually downdrafts more gentle and more extensive than the updrafts that they balance.

There is a rough but useful relationship between 'ceiling,' or height of cloud base, and surface temperature and dewpoint. At moderate temperatures, each degree of temperature-dewpoint difference means about two hundred and fifty feet of ceiling. Thus if the surface temperature is 70°F and the surface dewpoint 66°F, the ceiling should be around one thousand feet.

Occasionally the mechanism of updraft-builds-cloud is revealed in detached and clearly apparent form by a large fire. The smoke from the fire rises, of course, along with a great column of heated air. Just as the smoke begins to thin out, perhaps, at some moderate elevation, a new and thicker turruted murkiness appears and builds upward. This upper murkiness is not smoke, but cloud formed by the updraft, and the elevation where it appears is the condensation level. If a large and deep air layer is sufficiently unstable and sufficiently moist—conditions that exist in tropical or summertime thunderstorm weather—both the updraft and the consequent instability cloud may grow out of all proportion until the final result is vertical currents ballooning upward at possibly one or two hundred miles an hour, and a towering cloud mountain as high and wide as Everest.

For the wholesale condensation of water vapour into cloud droplets releases wholesale latent heat, which intensifies the updraft, which condenses still more moisture and releases still more latent heat, and so on. You can sometimes watch these thunderheads visibly growing against the blue sky— the faster they grow, the stronger the updraft and the more unstable the air to begin with. The upward momentum of rising air also helps in this cloudbuilding, which is ultimately checked, in extreme cases, by the great upper temperature inversion that is the stratosphere.

Instability clouds are not limited to hot weather. It is all a matter of relative warmth and coldness alow and aloft—of lapsed rate, provided enough moisture for cloud-building is available. A strong wintertime outbreak of frigid polar air over a relatively warm water surface (say along the Atlantic coast of the United States, or across the Great Lakes) can produce as puffy and squally clouds as anyone could wish. And in winter, with the sun low in the sky, these instability clouds and snow squalls look blacker and more forbidding than most summer storms.

Combined Cloud Formation

Not all clouds are distinctly of layer-like, quiescent stratus type or indisputably of puffy, instability cumulus type. It is possible to have both types present in the same section of sky, perhaps at different levels. Hybrid or combination types, such as strato-cumulus, are also possible.

The kind and prodigality of cloud-building depends on the relative strength and interplay of the three great cloud-building forces—moisture, instability, and external lifting. Any moist air layer bordering on instability tends to become more unstable, the more it is lifted by external forces. Such a layer, externally lifted, may at first form quiescent stratus clouds which, as the lifting goes on, may bulge upward here and there into growing instability cumulus turrets. Nevertheless, every cloud that we see in the sky must result from some chance combination of the three cloud-building forces. The more unstable the air is, the more moist it is, and the more violently it is lifted by external forces, the greater will be the bad-weather effects in clouds, rain, snow, hail, and the like.

We have said earlier that radiation is usually an unimportant factor in cloud formation. But moist air absorbs and radiates heat better than dry air, and it is possible for a high, moist layer (having a relative humidity close to one hundred per cent) to radiate away enough heat at night so that it forms a layer of high stratus cloud. Possible—but not common.

This ever-changing interplay of cloud-building forces in Nature's air ocean produces what, at first sight, seems to be an endless and bewildering variety of cloud forms. But these cloud forms (or the more common ones, at any rate) can be grouped into a few types and classes that are quite easy to remember and identify. The most basic cloud distinction of all, between layer-like stratus and heaped-up cumulus, has indeed already been made.

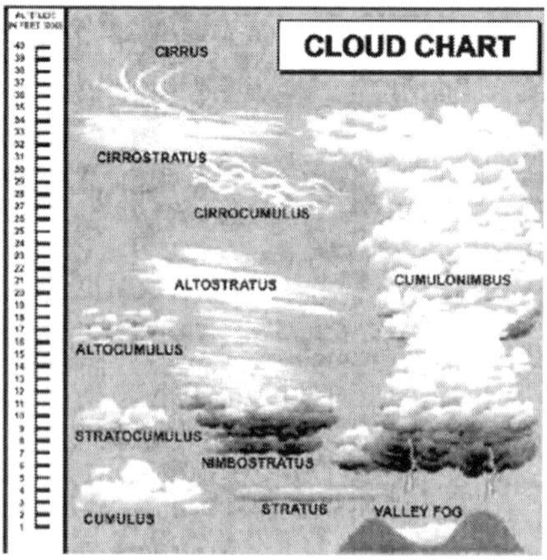

Fig. 3.4: Clouds by Type and Altitude

PARADE OF THE CLOUDS

The first attempt at naming clouds systematically and completely was made by Luke Howard, an Englishman, in 1803. He classified all clouds by shape or appearance into three general types—cirrus or 'curl' clouds, cumulus or 'heap' clouds, and stratus or 'layer' clouds. Other types of clouds, such as nimbus or rain clouds, Howard associated in some way with his three basic types. During the nineteenth century other cloud classifications were proposed, but Howard's scheme, from time to time elaborated and improved, gradually grew into a standard system increasingly accepted throughout the world. We are concerned only with the commoner clouds, such as any amateur can learn to name and recognize. And most amateur difficulties in the past have come from too-exhaustive or illogical systems of classification.

So we shall keep our cloud charts very simple, and as logical as possible. The most fundamental cloud differentiation, already hinted at in the preceding section, is based, like Howard's, on shape or form:

CLOUD CLASSIFICATION BY FORM OR SHAPES

Two Fundamental Shop, Classes

Layer class (from stable air)	Stratus (= 'a spreading out')	A quiescent cloud layer, rather flat and formless, extending horizontally for the most part.
part. Puffy class (from unstable air)	Cumulus (='a heap')	A piled-up heap of cloud, usually having definite threedimensional shape, often growing vertically.

Two Special Shape Types

Wispy type	Cirrus (= 'curl')	White, fibrous mares' tails or feathery curls at very high altitude.
Broken type	Such as Fracto-stratus * Fracto-cumulus	Heat-evaporated or wind-torn into separate smaller cloud masses with ragged, wispy edges.

Mixed and Storm Typer

Combination type	Strato-cumulus	Combining some charicteristics of two different classes.
Precipitation types	Nimbus (= 'rainstorm') Nimbo-stratus Cumulo-nimbus	Clouds of stratiform or cumuloform shape, respectively, from which rain, snow, or ice is actually failing.

Another way of classifying clouds, and perhaps a better one for understanding how they fit into the pageant of tropospheric weather, depends on their height above the earth's surface -or, in a more general sense, their altitude above sea level. Beginning with the highest and thinnest clouds, and ending with the lowest and thickest, the more common forms are:

Table. Cloud Classification By Altitude

Upper or citrus-like types	Cirrus Cirro-stratus Cirro-cumulus	Usually occur in the upper troposphere, often not far below the stratosphere, at heights of about 4 to 7 miles (6-10 km.).
Middle or high ('alto') types	Alto-stratus Alto-cumulus	Usually occur in the middle troposphere at heights of 2 to 4 miles (3-6 km.).
Lower types	Stratus Cumulus Strato-cumulus Nimbo-stratus	Usually occur in the lower troposphere, from the surface up to a height of nearly 2 miles (3 km.).
Vertical growth type	Cumulo-nimbus (thunderstorm)	Single cloud mass may extend from near the surface up to heights of 4 to 7 miles.

This table shows at a glance how various types of clouds fit into the general picture of the troposphere. From it one can appreciate why towering cumulonimbus clouds can be seen hundreds of miles away in clear air; one can sense why a spreading high deck of cirro-cumulus stands out in the red light of a clear sunset for perhaps three hundred miles to westward, where the sunrays, receding at ten miles per minute, still light the cloud deck's edges half an hour after they have left the ceiling directly overhead.

These high decks of upper clouds (cirro-cumulus or alto-cumulus), together with clear lower air to westward, are the one unfailing recipe for sunset splendor—were long before Meredith, and always will be: 'Ay, but shows the southwest a ripple-feathered bosom, Blown to silver while the clouds are shaken and ascend, Scaling the mid-heavens as they stream, there comes a sunset Rich, deep like love in beauty without end.'

What do all these cloud forms really look like, and how do they arise out of the vast emptiness of the air ocean? The cloud photographs. And in further elaboration, we review briefly the various outstanding peculiarities of each cloud form in turn:

CIRRUS

True cirrus clouds are fibrous and delicate in appearance, showing as white wisps against the upper blue. Sometimes they arc combed out into mares' tails; sometimes they are formed more thickly into curls on feathery plumes or parallel 'plowed-field' bands. But always they are the most tenuous clouds of the sky, scarcely dimming the sun's disk, and the highest, floating at five to seven miles above the earth (8 to 11 kilometers). As the temperatures at these levels are always far below zero, cirrus clouds are formed from minute, freefloating crystals of ice (simple hexagonal prisms). In fair weather they may result from general lifting of the upper troposphere above the active, 'colddome' portion of a high-pressure area. As foul-weather harbingers they are 'locks of the approaching storm,' blown out by substratospheric gales far in advance of the general cloud area of a cyclonic warm front.

Sometimes broken or wispy clouds fracto-stratus or (fractocumulus) in the lower levels have a superficial resemblance to cirrus clouds, and are mistaken for them by laymen. But the true cirrus clouds, carefully watched, are very evidently much higher than these fracto-forms.

Cirrus clouds are always interesting to watch because they show the motion of air near the top of the troposphere. Sometimes the long, fibrous tails seem to radiate from a cyclone centre below the horizon, and actually are combed out along the direction of air motion by the upper winds. At other times cirrus plumes or rays can be seen to move, not along their length but across it, like successive ranks of marching soldiers. Here the cirrus plumes make visible a colossal wave motion in the substratosphere—white crests of upsurging condensation and blue troughs of downdropping evaporation— great, silent aerial surges that may measure ten or twenty miles from crest to crest. Some cirrus clouds, again, are probably nothing more than evaporating wisps of what was originally cirro-stratus -a sort of 'fracto-cirro-stratus,' one might say. After sunset or before sunrise, cirrus clouds often glow with beautiful tints of flame-yellow and rose-pink.

Cirro-stratus

Cirrus-like cloud smoothed into one continuous, thin layer, resembling a whitish veil, is called ciffo-stratus. It may be smooth and milky, or again uneven and somewhat fibrous, forming out of thickening cirrus. Cirro-stratus never casts shadows, nor does it greatly dim either the sun or the moon, though in thickening and lowering form (approaching alto-stratus) it begins to blur their disks somewhat. Cirro-stratus is almost as high as cirrus (3½ to 7 miles, or 5½ to 11 kilometers), and is likewise formed of minute hexagonal ice crystals. In all citrus-like cloud-forms, but most noticeably in cirro-stratus, these ice crystals may refract and reflect light in such a way as to form halos or luminous rings around the sun or the moon. The most common halo is a single ring of about twenty-two degrees radius, coloured faintly if at all; but one or even two additional outer rings may be seen occasionally, together with rarer 'mock suns,' 'sun pillars,' tangent arcs, and the like.

('Coronas,' or rare, coloured sun-andmoon rings that may superficially resemble halos, are usually caused by water-droplet clouds at much lower altitudes.) Cirro-stratus cloud is most frequently formed in the upper and outer warm-front portions of a cyclonic cloud area, thickening to alto-stratus towards the storm centre, thinning out to true cirrus wisps in advance. Hence while quiescent detached cirrus is a frequent fair-weather cloud, cirrus thickening into a cirro-stratus veil is a more definite portent of possible cyclonic approach. The portent is still more definite if cirro-stratus thickens to altostratus— but by that time, the rain is not far behind.

Cirro-cumulus

Cirro-cumulus clouds are small, fleecy, white flakes or very small cloud balls—a 'mackerel sky'—perhaps merging into cirrus or cirro-stratus, and composed of supercooled droplets or ice crystals floating at great heights (3½ to 5½ miles or 5½ to 9 kilometers). True cirro-cumulus clouds are not very common, and are easily confused with thinning forms of alto-cumulus—which latter are always well below the true citrus-like cloud levels. The distinguishing feature is, however, that in cirro-cumulus the individual flakes usually appear to be smaller by reason of their great height, and they are never thick enough to cast shadows or show dark undersides.

Cirro-cumulus clouds may sometimes be formed by small up-and-down currents along a wide but shallow unstable layer in the upper air. But more often they are arranged, like some types of cirrus, in parallel rows or lines, showing wave-motion between overrunning and underrunning air currents somewhat similar to ocean waves along the sea surface where water meets differently moving air. All these high, cirriform clouds can best be seen, incidentally, through dark (preferably purplish) glasses; and they can best be photographed, like most other clouds, through red filters.

Alto-stratus

This cloud usually takes the form of a thick grayish or bluish veil, always somewhat fibrous in texture. It is thicker and less translucent than cirro-stratus, which often merges gradually into it with the approach of a cyclonic storm. Through typical alto-stratus the sun and the moon gleam only weakly and vaguely, as through ground glass; they have a watery and diffused look, or may be almost wholly obscured, usually casting no shadows. Unlike cirriform clouds, altostratus is composed mostly of minute water droplets plus ice crystals scattered here and there.

The usual height of alto-stratus is 2 to 3½ miles (3 to 6 kilometers). It occurs most frequently along the slightly sloping surface of a warm front, where the overrunning warmer, moister air, ascending the inclined plane of the cold-air wedge, has lost by condensation a good deal of its original warmth and moisture. This frontal alto-stratus often marks a zone where light rain or snow is just beginning to fall from the high clouds, though this outermost precipitation (except cold-weather snow) may evaporate again before reaching the ground. Aside from the typical warm-front cloud-sheet, local alto-stratus of a more level and quiescent type can occasionally result from the radiational cooling of a high and very humid air layer.

Alto-cumulus

Alto-cumulus is quite similar in general appearance to cirro-cumulus, but the individual flakes or cloud-balls of alto-cumulus are always larger and grayer. Usually they cast shadows and show dark undersides. The lower the alto-cumulus is, the larger are its opaque lumps. Certain forms of altocumulus ('castellatus') display little cloud-turrets growing upward from the main moisture-supplying layer.

In common with alto-stratus and most of the lower clouds, alto-cumulus is formed of minute water droplets, which may, however, be sub-cooled well below freezing in the liquid state. The usual height of alto-cumulus is 2 to 3 miles (3 to 5 kilometers). Like cirro-cumulus, alto-cumulus is often arranged in parallel rows along a wide and level deck, resulting from wave motion between differently moving layers of air. Alto-cumulus usually results from upper-air wave motion of some sort, or instability; it may grow into alto-stratus or nimbo-stratus.

Stratus

Any flat, shapeless, more or less uniform layer of cloud in the lower levels, usually dull gray in colour when viewed from below and whitish when viewed from above, is called stratus. Stratus resting on the ground is simply fog, formed and dissipated in the same way as fog; and high fog is simply low stratus. More usually, however, stratus occurs at heights of 1500 to 3000 feet (500 to 1000 meters). Low stratus is often a local cloud, gradually breaking away into shreds and smaller masses called fracto-stratus or 'scud.' Stratus can produce a fine drizzle, but not appreciable rain or snow.

CUMULUS

Cumulus clouds are thick, dense, detached 'woolpacks' or 'puffballs,' usually as high as they are wide. Their bases, as we have seen, are usually flat and uniformly fixed at the condensation level, but their upper parts are domeshaped and variously bulbed and rounded, like cauliflowers. Their colour ranges through every nuance of sunlight and shadow from dazzling white above to dark gray below.

It will bear repeating that a detached cumulus cloud always marks the top of a rising air current, which perhaps originated in the first place from an overheated portion of the earth's surface, or from some updraft in the general air flow. In any growing cumulus cloud conditional instability is being realized and condensation of moisture is contributing its latent heat to make the updrafts more violent. If moisture is plentiful and the instability extends upward to great heights, the cumulus grows into a cumulo-nimbus, or thunderstorm. But if an ordinary growing cumulus bumps into a very stable layer such as a deep temperature inversion, its top flattens and its growth stops.

Most ordinary fair-weather cumulus clouds, formed under clear skies at levels between 2000 and 8000 feet (600 and 2500 meters), wax and wane daily with the sun (or with surface temperature). Foul-weather cumulus clouds, on the other hand, occur at any time of day along cold fronts and within moist, unstable air masses. A torn, immature, or partly evaporated cumulus is sometimes called fracto-cumulus. Sometimes cumulus clouds on the lee side of a mountain range evaporate in the dynamically heated downdraft so rapidly that they seem to disappear as if by magic.

Strato-cumulus

Strato-cumulus usually occurs in the form of a wide and fairly level layer marked by thick rolls and dark, rounded masses of cloud. It somewhat resembles alto-cumulus, but its cloud masses are usually larger, thicker, and darker. Sometimes there are small or large open breaks between the rolls and masses; sometimes the layer is continuous but steeply undulating, especially along its upper surface as seen from an airplane. Strato-cumulus clouds usually occur in the same levels as fair-weather cumulus—say 3000 to 8000 feet (10002500 meters)—and may be formed by the flattening of cumulus clouds, instability in a stratus layer, or wave motion between two differently flowing layers of air. Strato-cumulus can change into, or develop from, stratus.

This cloud type, formerly called plain 'nimbus,' is in general any low, thick, formless, dark-gray cloud layer from which rain, snow or other precipitation is actually falling. Even if the rain momentarily stops, this cloud has a ragged 'rainy' look, and is still called 'nimbo-stratus.' Below the main cloud deck there may be various ragged 'scud' clouds (fracto-stratus or fractocumulus) that occasionally merge into nearly solid lower layers or into the main cloud deck above. Nimbo-stratus transmits a feeble illumination that seems to come from within the cloud itself.

Usually nimbo-stratus develops out of alto-stratus and rainformed lower fracto-stratus or 'scud,' marking the near approach of a warm front in the central region of a cyclone. Less frequently, nimbo-stratus may develop out of heavy and threatening strato-cumulus. In either case the nimbo-stratus cloud, at heights of perhaps 1000 to 5000 feet (300 to 1500 meters), results from widespread condensation in air forced to rise by reason of its movement around or towards the centre of a cyclone. Scattered areas of nimbo-stratus may develop from cumulonimbus, as in the rear parts of a thunderstorm.

Cumulo-nimbus

This is a great cumulus cloud mass in which condensation of moisture, release of latent heat, and increasing updrafts (fed by a deep layer of moist and unstable air) have grown to gigantic proportions. Its colour varies all the way from alabaster whiteness on the upper turrets to dark gray, black, or dark greenish below. The whole cloud mountain may tower from some ordinary cumulus level, such as half a mile, up to a height of five, six or even seven miles.

In a broad sense any large cumulus mass producing a shower can be called cumulonimbus, and this type is the common shower cloud, with or without thunder, of the tropics. But in temperate latitudes, cumulo-nimbus connotes more specifically a large, towering thunderstorm cloud whose top, blown out into anvil shape and cirrus-like texture by substratospheric winds, consists mostly of ice crystals. Hoods of alto-stratus may mark temperature inversions that the growing cloud mountain breaks through.

Special Cloud Forms

One special cloud form worth mentioning is mammato-cumulus or 'rain balls,' whose cumulus heads bulge downward instead of upward from a high cloud deck. It usually appears as a multitude of small, dark downward bulges in a turbulent layer of alto-stratus, alto-cumulus, or stratocumulus. The pendulous cloud pocks are evidently caused by local cold downdrafts perhaps started by detached showers of falling hail or snow. Mammato-cumulus always indicates very disturbed and unstable conditions in the upper air, seldom occurs far from a thunderstorm, and sometimes precedes tornadoes.

Another interesting cloud form is the detached, lens-shaped ovals that are sometimes formed at considerable heights out of quiescent or slowly dissolving cloud masses (alto-stratus 'lenticalus,' and so on). Sometimes the crest cloud on a mountain summit, caused by the steady flow of air up over the peak, takes this lenticular shape. The lens shape is produced, in any case, by symmetrical air flow and selective evaporation.

When high winds are blowing in mountain country, any tall detached peak is likely to fly a 'banner cloud' that streams out, pennant-like, far to leeward. Sometimes the great plume is formed of blowing snow from the peak itself. But if the humidity is just right, a true banner cloud can form on a perfectly bare peak such as Yosemite's Half Dome in summer, caused by the dynamic cooling effect of lowered pressure in the lee of the peak.

RAIN AND SNOW AND ICE

At ordinary temperatures all lower clouds are composed of minute water droplets perhaps 1/2500 of an inch across. Left to themselves in still air, these cloud droplets would sink slowly, at a speed less than one mile per hour. Yet as the droplets jostled together, somewhat larger droplets might be formed here and there. From a cloud mass (low stratus or high fog) near the ground in very moist air, these larger droplets, perhaps ten times as large as the original cloud droplets, might sprinkle downward in the form of a gentle drizzle.

But such a drizzle does not even approach the wetting power of real rain. Ordinary raindrops are ten times as big again as the drizzle droplets say 1/25 of an inch across. And rain clouds, considerably less common than ordinary clouds, are distinguished from them by two characteristics: fairly strong rising currents, and considerable depth of cloud. Many rain clouds indeed probably extend, summer or winter, well above the freezing level, and ice particles probably play an important part in their mechanism.

In any case the cloud droplets of any active rain cloud are being borne steadily upward, though not quite as fast as the air surrounding them. In the upper parts of the cloud the droplets are probably farther apart, and somewhat larger by reason of longer and freer condensation. Here and there they are combining into still larger drizzle droplets. Well up in the clouds the rising currents, not very uniform anyway, must somewhere begin to diminish.

As a result the larger droplets begin to fall downward, or at least rise slower than most of the cloud droplets, and in either case begin to sweep through the smaller droplets and the moist air surrounding them, gathering unto themselves more and more moisture as they fall. Finally these growing drops perhaps become large enough to continue earthward, even through moderate rising currents, in the form of rain. Much tropical rain and some temperate-zone rain probably originates in this simple way.

But in temperate and arctic latitudes probably most rain, and perhaps all heavy rain and certainly all snow, evolve originally out of ice particles. These ice particles can sublime directly out of water vapour onto sublimation nuclei (dust particles and the like, always present in the atmosphere) very easily at low temperatures around 5°F (-15°C) or colder; or, perhaps less frequently, they can be frozen out of ascending water (cloud) droplets that are likewise cooled well below freezing and jostled sufficiently by turbulence.

The level at which the ice particles are first formed varies with latitude and the season and the general weather situation from as low as a mile or less in winter to as high as four miles or more in summer. The upper parts of a typical heavy rainstorm, therefore, are mostly iceparticle clouds cirro-stratus, alto-stratus, and the like rather thin-textured clouds with indistinct edges. The lower parts of the same typical rainstorm, at least during its earlier phases, are mostly water-droplet clouds cumulus, stratus, and the like opaque-looking and well defined.

As the ice particles from the upper clouds fall slowly (at perhaps two or three miles an hour) through the lower (waterdroplet) clouds, a curious transfer of moisture occurs. By reason of physical laws that need not be elaborated here, at certain temperatures and humidities water vapour is evaporated from liquid droplets, yet sublimed on solid particles. Thus the falling ice particles constantly gain in bulk at the expense of the water droplets around them, and with increasing size fall gradually faster.

And in addition to their sublimation-growth, they freeze to themselves any subcooled water droplets that they happen to hit. The ultimate form of the ice crystals depends on the humidity, and varies from simple ice needles to skeleton stars (snow) to snow balls (graupel). As these ice crystals fall, the ice-particle cloud itself thickens and lowers, perhaps encroaching on the waterdroplet clouds below—or possibly engulfing them entirely, as in cold-weather nimbo-stratus. But after falling below the freezing level (provided it is well up in the air), the ice crystals soon melt into raindrops.

This is the idealized picture. But in a violent rainstorm -and most particularly, in a turbulent thunderstorm—cloud droplets, ice particles, raindrops, and snowflakes may all be milling around in some parts of the cloud.

RAIN

In 'light' rain the drops are very small, averaging perhaps 1/50 of an inch in diameter. Ordinary or 'moderate' raindrops are about twice as big (1/25 inch). Drops of 'heavy' rain are larger still, perhaps 1/16 inch wide. The very largest raindrops in a cloudburst may 1/8 inch or more across. Raindrops splashing on the ground of course look much bigger than the effective spherical sizes just given, and even while falling through the air they are somewhat flattened out.

The largest possible raindrop size is about 1 inch, which falls through still air at about twenty miles per hour; any larger drop, falling faster, would be broken into smaller drops by the wind of its fall. Though all rain results from rising moist air, the gentleness and uniformity or violence and variability of the formative updrafts go far to determine what form the precipitation will take. Rain out of stable (stratus) cloud layers, gently rising, say, along a warm front, is light, steady, and continuous. But rain out of unstable cumulus along a cold front (where updrafts are patchy but powerful) is likely to be heavy, variable, and showery.

In the heaviest cloudbursts rain falls at the rate of three or four inches of solid water per hour. (This excessive rainfall rate cannot continue for long, but even an hour of it may cause floods.) A fall of nine or ten inches in a few hours, by no means rare, brings creeks and rivers raging out of their banks and inundates the countryside. The heaviest rainfall so far ever recorded was at Baguio, Philippine Islands, where forty-six inches—nearly four feet of solid water on a level surface—fell in twenty-four hours!

Rain is most beautiful when it falls through clear lower air in the nearlevel rays of a low sun, glistening and sparkling in all directions, and perhaps painting the spectral arch of the rainbow. The primary bow, and perhaps a secondary are, are always centered about the observer's shadow. Hence they can never form more than half a complete circle for anyone on level ground or at sea. The cause of the rainbow is selective reflection and refraction of the various colours or wavelengths composing white light (at certain critical angles that vary slightly with the wavelength) by the small crystal-clear spheres of the falling raindrops. The primary bow, ranging from a sharp outer edge of red to a diffused band of inner blue or violet, has a radius of about forty degrees from the observer's shadow. The rarer secondary bow, if visible at all, begins rather sharply with red about nine degrees outside the primary and shades away outward into a nebulous greenish or bluish glow.

The larger the raindrops and the clearer the lower air, the more striking and beautiful are the rainbow colours. Ground observers, of course, see the largest bows when the sun is low. But from a high mountain-peak, or better still from an airplane flying between or below showery clouds, it is sometimes possible under a high sun to see a strangely unfamiliar but doubly miraculous and brilliant rainbow rounding into more than half a circle and seemingly laid on its side.

Less spectacular and less rare are the faintly coloured aureoles, called 'glories' or 'mountain-bows,' that anyone on a sharp peak or in an airplane can see around his shadow on the white surface of a sunlit cloudbank. If the cloud droplets are near at hand, these glories may resemble faint but orthodox rainbows. Smaller aureoles that appear on more distant cloud-floors—faint colour-rings only two degrees or so in diameter—are probably due to selective diffraction of light rays around the sharp opaque edge of the peak or airplane.

SNOW

Few places on the earth are not visited by rain, occasional or frequent, light or heavy, steady or showery. Over the whole earth, indeed, it has been estimated that rain is always falling at the stupendous rate of sixteen million tons per second! But in northern latitudes, through a large part of the year, this never-ending deluge takes the form of feathery-white showers of snow— snow that, unlike rain, accumulates everywhere, in the valleys and on the hills, to store moisture against a future melting. This snow blanket brings to sight general whiteness, and to sound general silence. Snow only four inches deep absorbs something like nine tenths of ordinary sound energy, being more efficient as a noise muffler than heavy draperies of velvet. A feather-like snowflake floating slowly downward seems to have little in common with a piece of ice. Nevertheless, that is what the snowflake really is, a piece of ice usually crystallized into a very intricate and beautiful pattern. Snowflake patterns are so elaborate, consisting mainly of outgrowths at sixty-degree angles, that each flake presents a large air-catching surface compared with its weight, and falls slowly like a parachute. Snowflakes can be caught on a dark cloth and examined under a magnifying glass, and they have often been photographed with microscopic cameras.

Snow, as already indicated, is probably formed by the same general causative factors as rain, operating however at temperatures entirely below freezing. As the nuclear ice crystals fall, they freeze to themselves as crystalline extensions any subcooled minute water droplets in their path, and grow also by direct sublimation out of invisible water vapour.

If snowflakes fall into temperatures above freezing on their downward path, however, each elaborate crystal melts into a drop of rain. In a climbing airplane it is not uncommon to sound through a region of rain up into a zone of snow (though the exploration is not particularly to be recommended, owing to icing hazards), and much winter rain undoubtedly begins as snow in the higher levels.

Although snow can melt into rain, rain cannot freeze back into snow. If rain from an upper, warmer air layer passes through a colder-than-freezing layer close to the ground, the result is frozen raindrops, or round pellets of clear ice, properly called sleet. Sometimes, however, snow falls in the form of round, compacted pellets which resemble sleet, but are more correctly called graupel; and occasionally these pellets come through to the ground. The remaining form of aerial ice is hail -pellets formed by repeated up-and-down rides in the towering thunderstorm clouds.

If rain falls in liquid drops on ground, houses, trees, or wires that have been previously cooled below freezing, it forms a layer of clear ice called glaze. 'Icing' of a somewhat similar sort can occur (to the embarrassment of pilots) on the wings or other parts of an airplane in flight through cold clouds or rain. Another type of ground icing (also noticed on airplanes) is rime, where subcooled fog or cloud droplets freeze on exposed projections, gradually building out into the wind. Widespread and continued glaze, which may bring down trees limbs, telephone wires, and power lines by the hundreds, is called an 'ice storm.'

The formation of clouds out of thin air, the birth of precipitation out of clouds—both these processes are physical experiments conducted by Nature in her open-air laboratory on a gigantic scale. Even a local summer shower involves millions of tons of air, hundreds of thousands of tons of water, and energy equivalent to the burning of thousands of tons of coal. A widespread storm, such as occurs along the path of a cyclone, may involve materials and forces many times greater.

❐

6

Wind and Climate Belts

This great, rotating ball that is Earth harbours within its atmosphere a thousand varieties of wind and weather over the diverse oceans and continents of its expansive surface, even at any one instant of time; and all these weather vagaries are constantly changing, quickly or slowly, from day to day, week to week, month to month.

Nevertheless there is a grand and logical scheme underlying general world-wide weather—a pattern that unmistakably appears in the average weather, or climate, experienced at representative places spaced up and down the meridians of latitude; and in the prevailing winds that arise out of these belt-like, world-encircling regions of climatic unlikeness.

This average planetary wind-and-climate picture is seldom, if ever, duplicated by actual world-wide weather on any single day—particularly in north-temperate regions, where actual weather variability seems to discredit the existence of any such regular scheme. Yet over the oceans—where the sea surface is relatively flat and relatively frictionless to wandering breezes, where sameness of temperature discourages climatic oddities and local weather freaks, where the climate belts and prevailing winds are most evident, typical and regular—it is possible to glimpse in actual weather some prominent features of the grand planetary pattern. You can best sense certain of its gradual but definite and continuing changes, from latitude to latitude, by cruising north or south in a seagoing ship.

Here, for example, is the actual record of a March cruise from Panama to Philadelphia in the old but able American freighter Timber Rush—a slow twothousandmile cruise from torrid to temperate latitudes at the comfortable ocean-going dog-trot of ten knots:

March 11th: latitude 10°N, off Panama coast, temperature 79°F, broken clouds, wind ENE moderate (trade wind), caught barracuda on trolling line.

12th: Lat 12°, in Caribbean, temp 79°, scattered to broken clouds, wind NE X N fresh to strong (norther), sea rough, water temperature 78°.

13th: Lat 14½°, in Caribbean, temp 77°, scattered clouds, wind ENE fresh (intensified trade), water very rich blue.

14th: Lat 18°, just southwest of Haiti, temp 78°, scattered, clouds to clear, wind E'ly light (in lee of Hispaniola).

15th: Lat 22°, near Bahamas and Tropic of Cancer, temp 77°, scattered clouds, wind E'ly to variable light (horse latitudes) scattered clouds to clear.

16th: Lat 26½°, in Atlantic, temp 73°, showers and thundershowers along a cold front, wind SW'ly moderate shifting to NW fresh with frontal passage.

17th: Lat 30½°, in Atlantic, temp 67°, scattered to broken clouds, wind NW moderate (prevailing westerlies), water dark steely blue, water temperature 74°.

18th: Lat 35°, off Cape Hatteras in Gulf Stream, temp 60°, broken clouds, wind N'ly fresh and increasing, water temperature 72°, passed school of whales.

19th: Lat 38°, entering Delaware Bay, temp 35°, clear, calm to S'ly light, water temperature 50° after leaving Gulf Stream, 42° near coast, water dirty green.

Thus we passed, in the last four days of slow steaming at two hundred and forty easy sea miles a day, from a tropical climate to a temperate climate, from smiling summer skies and tufted cloudlets to cold, murky storms and driving showers, from moist, warm, enervating sameness to irritating and unexpected variability.

CLIMATES, TROPIC AND ARCTIC

Variability—that is the outstanding characteristic of temperate-zone weather. Even the brief sea log just quoted shows how weather in the tropics is much the same from place to place and day to day; temperate-zone weather, conversely, is diverse and changeable. But had our voyage extended far enough northward, it would have reached another region of comparative weather sameness—cold sameness instead of warm sameness -the arctic zone.

The temperate zone can have no stable and typical weather of its own. To southward (in the northern hemisphere) is the steady, warm, moist climate of the tropics. To northward is the fairly steady, mostly cold, fairly dry climate of the arctic. (This picture is of course repeated, with reverse directions, in the southern hemisphere.) The temperate zone is a buffer region between tropic and arctic climates, a battleground on which the gigantic atmospheric forces of heat and moisture advance and retreat, an unstable compromise between irreconcilable extremes, swayed this way and that in its evanescent weather as tropic forces (warm air masses) or arctic forces (cold air masses) momentarily gain ascendancy.

Tropic Climate

Tropic wind-and-climate belts migrate somewhat, north and south, with the apparent north-south seasonal swing of the sun. To get within-the-tropics weather in its most typical form we must go well south of the Tropic of Cancer, nearly down to the equator—say to Panama in about latitude 8°N. In these latitudes there are only two main seasons a winter dry season under the influence of the steady northeast trades, and a summer wet season brooded over by the fitful calms of the doldrums.

The outstanding features of tropical-marine climate are that the average temperature is practically constant from month to month (nor does it ever vary, by more than a few degrees, from day to day); and that the cloudiness and rainfall, always appreciable, reach high values during the wet-season doldrums invasion of summer and early fall.

Upper winds over Panama, at half-mile altitude, average northeast about 8 m.p.h. in the rainy season and north-northeast about 20 m.p.h. in the dry. At two miles altitude there is less seasonal change—the average wind only varies from east 10 m.p.h. (rainy) to cast-northeast 12 m.p.h. (dry).

Upper-air characteristics over the Caribbean Sea, quite typical of tropicalmarine climate in general, were studied during the 1939 hurricane season by a United States Weather Bureau expedition to Swan Island. This four-milelong islet, named after a nineteenth-century pirate, lies almost exactly in the middle of the Sea at about lat. 16°N, between Jamaica and British Honduras. Within four months, observers Rhamlow and Leech released some two hundred and fifty pilot balloons and one hundred and twenty radio-sounding balloons. They found that the trade wind extended well aloft, though with diminishing force, here as in Panama—prevailing easterlies all the way up to about two and a half miles altitude. But above this level, the upper winds are mostly out of the west and increase with altitude.

Surface temperatures at Swan Island, and all over the Caribbean area, average about +75° to 80°F the year round. But at the base of the stratosphere, ten miles up in these latitudes, it is cold indeed—about—105° to—110°F. Thus in the thick tropical troposphere there is an average lapse rate of something like—3½°F/1000 ft (about—6°C/Km). Together with plenty of warmth and moisture in the lower levels, this lapse rate means that conditional instability must extend through considerable air layers, particularly in the vertical-sunwarmed rainy season of summer and fall—and it is this conditional instability, together with inexhaustible supplies of moisture from the warm ocean, that produces the daily torrential showers and thundershowers, the regular afternoon downpours and the moist, misty mornings which can make equatorial regions such a continual sweat-bath for ill-adapted white men. On or close to the ocean this deep-tropical weather is bearable, or perhaps even comfortable. But the steaming jungles inland, where temperatures can rise unchecked and where all breeze is choked off by the tangled uprearing of dense vegetation, are sometimes aptly described as 'green hells.' As Julian Duguid wrote ' a rich eternal garment of green, dappled with golden sunspots a dense, fever-stricken thicket shimmering in the heat with a perpetual glassy haze.'

However, by no means all the land areas within the tropics are covered with jungle—some of them are sandy and arid in the extreme, and doubly hot though not so muggy. (The word 'jungle' itself, originally, meant only 'waste land' in India.) And even in the fecund jungle areas of the tropics, such as Panama and northern South America, vegetation waxes lush and green with the rainy season, wanes sparse and tawny with the dry season.

There are also irregular year-to-year variations in moisture and prevailing winds that change the face of tropical nature. I got off a ship on the Colombian coast in August, 1939, and walked inland for a while, expecting to wade through wetseason green tangles of vines and trees peopled with tropic creatures—but the soil proved to be dry and sandy, the jungle a dried-up shrinkage of scrub, the creatures hidden away from the glaring vertical sun.

Arctic Climate

Before Admiral Peary of the United States Navy marched five hundred icy miles on snowshoes from Ellesmere Land to the North Pole in the early spring of 1909, and since his time, many expeditions have entered the northern arctic regions. No latter-day explorer, incidentally, has ever surpassed Peary's record of intelligent planning and courageous, competent performance. But the airplane has rung startling changes in arctic travel.

In May, 1937, four large air transports set down four Russian scientists and a dog, together with a little portable cabin and food and fuel for a year or more, on one of the uncertain drifting ice floes that cover the Arctic Ocean around the pole. Leader Papanin, astronomer-aerologist Federov, hydrologist Shirsov, and radioman Krenkel had hoped to remain at or near the earth's axis for a whole year of observations, but their ten-foot-thick ice floe soon began to drift at first very slowly, across the two-mile-deep polar basin towards Greenland. Then it moved, less gradually, southward along the east coast of that high continent, and finally sailed at twenty-five miles a day on southward.

Papanin's expedition found the summer temperatures of the polar basin hovering mostly around + 30° to 46°F; clear spells of a day or two alternated with dull, overcast, rainy-to-wetsnowy weather. Winds varied from calm to light or moderate, mostly out of the north or northeast. Much of the ice was melting in the warmer spells, adding to the troubles and perils of the icecampers. There was always the danger that their floe would disintegrate and dump them unceremoniously into the Arctic Ocean.

When winter came they had drifted several hundred miles south of the pole itself, but winter climate around the polar basin is probably pretty much the same everywhere. Temperatures ranged all the way from 50°F below zero to 30° or more above, in accordance with the sweep of unlike air masses and cyclones, which occasionally extends well into the Arctic. For several months in midwinter the polar region is, of course, in continual darkness, with the sun visible at first in the spring only as a pale horizon glow around noon. On the thirty-first of December, strangely enough, the Russian wind-powered radio reported a temperature of + 43°F over the arctic ice! But within a few hours, the mercury had tobogganed to a more usual value of -31°F.

The general north-polar region is, of course, about midway on the most direct (great-circle) air routes between Europe and Asia, and between eastern United States and Asia. Some years hence, perhaps, great 500-mile-an-hour strato-transports will be winging daily across these icy-watery wastes. Compared with other climatic regions, the polar area offers good flying conditions—during at least two or three days out of a week, that is, and particularly at higher levels. Spring, when the polar high-pressure area lies symmetrically over the geographic pole, and not over the continental 'cold poles' in Greenland and northern Siberia, is the best flying season. A sealedcabin, supercharged stratoliner, in any case, would be well up in the low arctic stratosphere (above thirty thousand feet) most of the time.

Aside from Ellesmere Land at the north tip of America, one of the best fixed bases for observing the climate of the north-polar basin is the small island of Franz Joseph Land in lat. 81°N long. 60°E, about six hundred land miles from the pole. In 1932-33 the Russian scientist Moltchanov (incidentally one of the pioneers in the development of radio-sounding balloons) made both surface and upper-air observations on Franz Joseph Land for a year.

Like the floe-campers, he found average summer surface temperatures around +30° to 40°F. Winter temperatures averaged around -10 F. Above the ground, for the first mile, there was usually a temperature inversion amounting to 5°F or so; above that height the lapse rate or vertical temperature gradient was always surprisingly uniform at -3° or -4°F per thousand feet up to the low polar 'tropopause,' marking the beginning of the stratosphere at about five miles altitude in winter, six miles in summer. Temperatures at the stratosphere base averaged about -75°F in winter, about -40°F in summer. July and August were the warmest months, as usual in the northern hemisphere; December was the coldest month at the surface, February coldest aloft. The southern hemisphere in general, being mostly water, shows more typical wind-and climate belts than the northern hemisphere, which is all cluttered up with land masses that get too warm in summer and too cold in winter. By the same token, or a reverse one, the antarctic climate, founded upon a high, icecovered continent symmetric around the geographic pole, is more typically 'polar' than the arctic climate.

During the 1933-35 Byrd occupancy of 'Little America' in about lat. 79°S, long. 164°W, aerologists Haines and Grimminger recorded a great deal of antarctic weather. Mostly the average temperature was around + 20° to 25°F in summer (January), and around -35° to -40° in winter (July) considerably colder, that is, than the north-polar climate. The coldest summer temperature was +12° on January 1; the warmest, +37° on December 22. In winter the mildest reading was +23° on July 27; the coldest was -71° on July 21. During a roaring blizzard in July the temperature fell to -65° with the wind sixty miles an hour out of the southeast and wind, rather than temperature, is what really hurts in cold climates. These southeasterly blizzards in Antarctica correspond to the northeasterly gales around the north-polar basin.

High winds in Antarctica sometimes come from the north, just as northpolar gales can arise out of the south, but the colder blizzards from somewhere around the South Pole are more usual. Laurence M. Gould, Byrd's geologist and deputy in 1928-30, writes of one terrific southerly gale in the Rockefeller Mountains that first blew away the heavily packed snow wall in front of his firmly anchored airplane, and then finally blew the plane itself away and wrecked it. Some time before the maximum wind came, the airplane's airspeed meter indicated a steady velocity of ninety miles per hour, with gusts well over a hundred.

Conquest of Climate

The decade from 1900 to 1910 signalized an. amazing conquest of the tropics as a livable environment for civilized man. The years after 1940, likewise, may see a parallel conquest of the arctics. Within the tropics, except in certain areas plagued by extreme and wearisome heat, general living conditions were easy enough too easy for man's own good, in fact; and aside from air transport, radio communication, mechanical refrigeration, and other new conveniences, the conquest was nearly all medical.

Such dread warm-country diseases as typhoid, cholera, and yellow fever were finally vanquished at least in the case of selected, well policed areas. The work of Walter Reed, his assistant surgeons, and the heroic American soldiers who voluntarily entered the valley of 'yellow jack's' macabre shadow -all this really conquered the tropics and equatorial regions for white men, and contributed more than engineering and steam shovels to the triumphant completion, in 1913, of the world's greatest inter-ocean canal. Within the arctics and around the poles, on the contrary, disease is unknown unless brought to the scene by human carriers, but living conditions (without many bothersome precautions) are unduly hard and exacting. Aside from the extremes of dark and cold and gale and blinding snow, the chief hazards are: difficulty in finding your way around (particularly in the long, cold, dark arctic night); difficulty in surviving ordinary mischances such as falling into water, badly tearing clothes, and so on, particularly if alone; difficulty in getting adequately balanced diet and sufficient ultra-violet radiation.

Yet even now, with our immense progress in transportation and communication, some of these difficulties can be largely overcome at a properly organized arctic station. Clothing must, of course, be of arctic type—layers of wool or fur inside for warmth, layers of canvas or leather outside for windbreaking. Adequate shelter, plenty of fuel, and fresh food can now be transported easily by airplane to almost any arctic site. On location, howling arctic gales will provide plenty of electric power for lighting and radio communication, and perhaps for heating as well.

Many of the dark terrors of the outer arctic night, which have heretofore held men prisoned for months in their base camps (except in the full of the moon) can perhaps be conquered by such devices as more powerful and efficient portable searchlights (using super-hot filaments, say, or possibly fluorescent light-sources); portable radio-telephone sets for constant two-way communication with and direction-finding on one or more base stations; more compact and accurate instruments, and simpler methods, for celestial navigation. Against the white, too-uniform snow-glare of the polar day, improved goggles (perhaps of polarizing type) may be employed.

Lighter, stronger and warmer tents and clothing can perhaps be devised. Much of the heavy local leg-work of arctic exploration can now be done with snow-mobiles, either tread-driven or propeller-driven. By means of vitamin diets, ultra-violet sunlamps and radio-diathermy, human beings near the poles will perhaps be enabled to maintain a high state of health during the long arctic night, and to overcome quickly any respiratory diseases that do develop.

TEMPERATE CLIMATE

We have said that the temperate zone is, in a climatic sense, nothing better than a battleground between tropic and arctic weather forces—between warm and cold air masses. Nevertheless the temperate zone (and in particular, the north-temperate zone) is that part of the earth in which civilized man evolved, and is probably the region most favourable to his continued progress and development.

On the average, of course, temperate temperatures are pleasantest and best for all human activities—neither too hot nor too cold—though this average may rarely be matched by the vagaries of daily weather. And even the extreme moodiness and changeability of temperate weather has its advantages. Variety in weather, at least as to the four- or five-day spells and sequences so common in middle latitudes, perhaps satisfies some deep, innate rhythmic urges in man which cannot be allowed to atrophy without intellectual and cultural and physical loss.

However such obscure causes may or may not operate, it is certain that most human progress up from the dumb-ape level—most human art and ingenuity and imagination, most human science and engineering and social organization—has been best nurtured so far, and probably will be best nurtured in the future, in temperate latitudes.

Among all the various meteorological factors that go to make up a climate, temperature is the one that most directly influences human health, happiness, and efficiency. How do various temperatures really feel to average human beings? Different degrees of heat of course feel quite different to different people, or to the same people under different conditions—in particular, under varying conditions of exertion. But it is possible to indicate average human reactions to temperature, and they are somewhat as follows:

Table. Outdoor Temperatures Feel to the Average Person

Temperature		Human Effects
(Fahrenheit)	(Centigrade)	
Below 0°	Below - 18°	Unpleasantly cold without heavy arctictype parkas, etc., unless exertion is continuous and heavy.
0°-32°	- 18°- 0°	Feels cold with anything but arctic-type clothing, if exertion is light and exposure long.
32°-50°	0°-10°	Under same conditions, feels comfortable with heavy overcoat, hat, etc.; cold otherwise.
50°-65°	10°-18°	Feels cool with ordinary indoor clothing, comfortable with light overcoat, etc.
65°-75°	18°-23°	Feels comfortable with ordinary clothing.
75°-90°	23°-32°	Feels warm co hot if humidity is high, wind light; feels comfortable in bathing suit.
90°-100°	32°-40°	Feels hot under any conditions, unpleasantly hot with high humidity and light wind.
Above 100°	Above 40°	Unpleasantly hot under all conditions.

High humidity makes warm temperatures feel warmer, by retarding evaporation from the body. High humidity also makes cold temperatures feel colder, by making the clothing conduct heat better away from the body.

Wind makes cold temperatures feel colder, by removing the cushion of warm air that tends to form around the body in the absence of wind. This wind-cooling effect is particularly powerful at below-zero temperatures. On the other hand, wind makes warm temperatures up to 100°F feel less warm, by increasing evaporation from the body. But if the temperature is over 105°, wind increases the discomfort. The optimum temperature range, 68°-70°F, feels more comfortable if the air is normally moist.

For health and comfort, indoor relative humidity should be around sixty per cent. But in the average home or office, in winter, it is seldom more than twenty per cent. Evaporating pans on radiators help some, but not very much. Constant-duty electric evaporators, preferably combined with electric lights, might well be used in temperate-zone houses during the winter.

Table. Indoor Temperatures feel to the Average Person

Temperature		Human Effects
(Fahrenheit)	(Centigrade)	
Below 64°	Below - 18°	Unpleasantly cool with ordinary clothing.
64°68°	18°-20°	Too cool for comfort if sitting still.
68°-70°	20°-21°	Optimum for health and comfort.
70°-74°	21°-23°	Warm but not uncomfortable if clothing is light.
74°-78°	23°-25°	Needlessly warm for healthy people.
Above 78°	Above 25°	Unpleasantly warm for healthy people.

Today's temperature anywhere is indicated by the thermometer; tomorrow's (which in temperate or arctic latitudes may be shockingly different) is foretold by a good forecast. Beyond these limits, our best present guide to future temperatures is the 'normal.' To get the normal temperature for any particular day of the year at any particular place, the average temperatures for that day, observed over as long a span of years as possible, are all averaged together. The normal temperature is, of course, merely a fictitious quantity seldom or never duplicated by any actual daily average; for climatically speaking, we dwellers in the temperate zone are always traveling now south, now north with the ebb and flow of warm and cold air masses.

Is our North American temperate-zone climate changing? This question keeps recurring in weather talk, and most of the oft-quoted answers foist off something about 'colder winters in Grandfather's day' a notion perhaps promoted by oldsters in support of their claims to general superiority. The rising generation, contrariwise, have been inclined to charge the whole idea off to dotage and general boastfulness. A definitive answer has been given by J. B. Kincer of the United States Weather Bureau:

For a good many years up through 1936 there has been a decided tendency to warmer and drier weather, the trends being especially marked during the past quarter of a century, notwithstanding an occasional bad flood or a severely cold winter. Take the winter season, for example: With the exception of those for 1917-18 and 1935-36, the winters for the past twenty-five years or more, considering the country as a whole, have been rather uniformly warmer than normal, and, on an annual basis, every year since 1929 has had above normal temperature. Also, in the matter of rainfall, there has been an equally marked tendency to droughts in recent years. However, an examination of the longer weather records of the country, going back one hundred years or more, indicates that this does not represent a permanent change of climate, but rather a warm, dry phase of our normal climate, to be followed, doubtless, by a cooler, wetter phase, when there will be more rain in summer and lower temperatures in winter.

Since the turn of the century, Mr. Kincer goes on to say, there has been a distinct yearly trend towards warmer winters and drier summers all over the world and this in spite of a few exceptionally cold winters here and there.

In other words, there are long (say fifty-year) cycles in climate, just as there are short (say five-day) cycles in weather. If all these cycles were only considerate enough to maintain constant length, amplitude, and form, both weather forecasting and climate forecasting would be child's play. But the form, intensity, and length of weather and climate changes are all complexly variable.

In any case the semipermanent (say 500-year) picture of our American climate shows, apparently, no appreciable change. But the geologic-time picture, measured in thousands or millions of years, would again show marked climatic cycles the advance and retreat of the various ice ages.

THE PLANETARY PATTERN OF WINDS

Just as there are on this great globe fairly definite zones of climate, so there are even more definite belts of prevailing wind. Some of these round-theworld prevailing winds are quite steady and uniform, such as the trades and anti-trades. Others, such as the prevailing westerlies of temperate latitudes, are extremely variable from day to day and week to week. But the average wind directions observed in temperate latitudes prove that here, too, there is a great round-the-world wind system from west to east—winds that blow generally across continents and oceans, across desert sands, sawtooth mountains, fields of rippling grain, waving trees in virgin forests, and wide expanses of heaving, restless, white-capped blue water. The winds of the world are in essence a heat-pressure mechan- ism similar to that which causes a sea breeze, though enormously larger and more complex. The driving force behind all earthly winds, and all other earthly motions, is the enormous outpouring of radiant energy from the sun, our celestial ancestor and gigantic powerhouse ninetythree million miles out in space.

This radiant energy falls nearvertically on equatorial regions, and very obliquely on polar regions. Hence the equatorial regions are warmer, and are receiving more heat than they lose by earth-radiation. The poles, on the other hand, are radiating away more heat than they receive by sun-radiation. Hence, unless the tropics are to become unbearably hot and the arctics are to chill down to absolute zero (about -460°F. or -273°C), there must be some atmospheric circulation between the two. A heated region usually develops low pressure at the surface and high pressure aloft, that a cooled region develops high pressure at the surface and low pressure aloft. The result, where only short distances are involved, is that lower winds blow generally from cold to heat, and upper winds blow generally from heat to cold. Hence we might expect, as part of the planetary wind picture, some tendency towards an equatorward drift at the surface together with a poleward drift aloft.

However, that any largescale wind is gradually turned aside, to the right in the northern hemisphere and to the left in the southern, by the earth's slow rotation under the wind in question. In other words this 'earth's deflective force,' so called, tends to turn any equatorward-drifting air westward, and to turn any poleward-drifting air eastward.

In the upper troposphere, say at altitudes above two and a half miles, the average world wind pattern is a grandly simple thing that stems directly and understandably out of these elementary heat-pressure-wind principles. Upper pressure is relatively high over the equator, and relatively low over the poles, with a fairly uniform average horizontal pressure gradient between the two.

Hence the average upper-tropospheric winds of the northern hemisphere, directly driven by this pressure gradient and swayed by the earth's deflective force, are from southerly to southwesterly to westerly (drifting, that is, generally towards the northeast, or spirally towards and around the pole). In the southern hemisphere, driven by similar but oppositely directed forces, the average upper-tropospheric winds are mostly northwesterlies, drifting generally southeastward.

Either of these average drifts can be temporarily upset by regional vagaries, such as a southward surge of cold upper air in the north hemisphere. But as a norm or average condition, the two great upper drifts perennially move from the equator spirally eastward and poleward in each hemisphere.

The planetary pattern of lower winds (below the two-and-ahalf-mile level) is more complex, even in idealized form. For the lower winds are influenced by surface friction, and the earth's deflective force acts on them as a more drastic limitation on poleward or equatorward progress. Thus the planetary lowerwind pattern includes, not one wind belt in each hemisphere, but three. Moreover, these wind belts, and the pressure belts which separate them and regulate them, are in fact much modified and distorted (at least in the northern hemisphere) by the alternating surfaces of great continents and great oceans.

Perhaps the easiest and best introduction to them would be a voyage through ideally typical weather at just 5° of latitude (three hundred sea miles) each day down the whole 'length' of the world, along the ocean-parting meridian of long. 30°W from arctic to antarctic. Any such imaginary meridianal voyage would cut across the following idealized or average wind-andweather belts: a zone of arctic easterlies above lat. 60°N a stormy 'polar frontal' zone (actually variable in both width and location) around lat. 60°N or south of this; prevailing southwesterlies-to-westerlies (the familiar prevailing winds of temperate latitudes) between lat. 60°N and lat. 30°N a dry, clear calm belt called the 'horse latitudes' (probably because horses often died here on becalmed sailing ships) around lat. 30°N northeast 'trades' (so called by reason of their steady, perennial character and consequent importance to commercial sailing ships) from south of lat. 30°N down towards the equator: a humid, oppressive, and showery calm belt called the 'doldrums' (because of their dispiriting or dulling or 'dolting' effect on human sailors unwillingly becalmed in them) around the equator; southeast trades from near the equator down to somewhere near lat. 30°S the southern horse latitudes around lat. 30°S; the southern prevailing westerlies between lat. 30°S and lat. 60°S; the southern polar-frontal zone around lat. 60°S or north of here; and the antarctic easterlies south of lat. 60°S.

The steady poleward air transfer of the prevailing westerlies and southwesterlies in our latitudes is compensated by occasional large and violent southward outbreaks of polar air (cold waves), especially in winter, when the polar anti-cyclone extends down over Alaska and northern Canada. Southward outbreaks sometimes also occur far aloft, with 100-m.p.h. north winds in the stratosphere. Similar equatorward (northward) outbreaks occur in the southern hemisphere.

Considering first the northern hemisphere: There is an average highpressure area, or anti-cyclone, around the 'cold pole' (despite the fact that many arctic observers have noted low pressures). This polar high-pressure area drives the cold, arctic lower air spirally outward (or southward and westward) as the arctic easterlies (mostly northeasterlies) down towards lat. 60°N. Along lat. 30°N (the horse latitudes) there is a subtropical high-pressure belt, the northern slope of which drives the prevailing westerlies (mostly southwesterlies) up towards lat. 60°N. Around lat. 60°N or south of there, these two great and conflicting wind systems meet along the low-pressure belt of the polar-frontal zone—a stormy region which may actually swing far north or south, and may extend as separate and vari-formed fronts, between various unlike air masses, through a wide sweep of latitude. The subtropical high-pressure belt in lat. 30°N (or its southern slope) also drives the northeasterly trades, as a general southwestward-to-westward drift around the earth, down into the equatorial low-pressure belt of the doldrums.

In the southern hemisphere, the subtropical high-pressure belt around lat. 30°S drives the southeast trades up into the doldrums, and also drives the southern prevailing westerlies (mostly northwesterlies) down into the lowpressure polar-frontal zone around lat. 60°S or north of there, where they meet and battle with the antarctic easterlies (mostly southeasterlies) from the southpolar high-pressure area. In temperate latitudes, both north and south, lower winds have the same general or average direction as upper winds, and the prevailing westerlies form a deep average pattern (seldom fully duplicated in actual existing winds) that extends with increasing force up through the troposphere, moderating again in the lower stratosphere, and then mostly reversing, in light to moderate intensity, above twelve miles or so. Within the tropics, on the 'winter side' of the equator, the easterly trades give way, at about two and a half miles altitude, to the planetary upper westerlies.

Called the counter-trades or anti-trades, these tropical upper winds blow mostly from southwesterly in the northern hemisphere and northwesterly in the southern. The same planetary upper westerlies are present, to some extent, in arctic regions above the arctic easterlies.

CENTERS OF ACTION

The whole idealized wind-belt picture described above is based, it will bear repeating, on average conditions. Daily departures from this average, in any one region, may be very wide. It is also based on the conception of a uniform earth—either all water, or all uniform land. As a matter of actual fact the southern hemisphere is so nearly all water that this idealized pattern actually holds there much of the time; and the pattern may also hold, though less consistently, well out in the Atlantic and Pacific Oceans of the northern hemisphere.

But in general the northern hemisphere is something like equally divided into great land surfaces and great water surfaces, which vary greatly in the frictional drag they oppose to surface winds.

Also, while oceans tend to remain equable in temperature at all hours and all seasons, continents tend to get too warm in summer and too cold in winter. As cold areas tend to develop high surface pressure, and warm areas low surface pressure, the idealized latitudinal pressure belts of the northern hemisphere are in fact somewhat broken up into continental and oceanic 'highs' and 'lows' which change considerably with the seasons.

These semipermanent, or seasonal, lows and highs are called centers of action, because their counter-clockwise or clockwise circulations control in some degree the general movements and conflicts of air masses in their neighbourhoods, and influence greatly both the position and movement of various segments of the polar front and the development and motion of the traveling cyclones that are largely responsible for most temperatezone bad weather.The centers of action in and around North America—those that have most bearing on our weather in the United States -are as follows:

Winter	Summer
Canadian High	Hawaiian High
Aleutian low	Bermuda-Azores high
Icelandic low	(Rokies low - weak)
	(Aleutian and Icelandic lows - weak)

It will be well to recall these centers of action, and to remember the cardinal fact that (in the northern hemisphere) winds blow clockwise about a highpressure area and counter-clockwise around a low-pressure area—particularly when we come to consider the traveling 'highs' that are usually associated with vagrant air masses, and the traveling 'lows' or cyclones out of which grow warm and cold fronts.

Table. Typical Climate of The Deep Tropics

Month	Mean Temperature	Mean Cloudiness (% of sky)	Monthly Rainfall (inches)	Average Wind direction and force
Jan.	80	50	3.5	North, * moderate
Feb.	80	52	1.5	North, moderate
Mar.	81	50	1.5	North, moderate
Apr.	81	61	4.2	North, light
May	81	75	12.6	Variable, light
June	81	81	13.7	Variable, light
July	81	81	15.6	Variable, light
Aug.	80	80	15.2	Variable, light
Sep.	81	77	12.5	Variable, light
Oct.	80	77	15.5	Variable, light
Nov.	79	79	22.2	Variable, light
Dec.	80	61	11.1	North, light

What are the outdoor conditions of life and thought and feeling near the poles and within the arctics? As a brief and partial answer to this always interesting question, Frederick Crockett, who was also on the 1928-30 Byrd expedition, gave me the following word-picture:

Usually during the months-long arctic 'day' there are low-hanging clouds, but below them, along the ground, visibility may be very good. The view in all directions is extraordinarily uniform—no shadows, no dark objects, no horizon, all gray above blending with chalky white below, 'like walking in a bowl of milk.' On this white, featureless surface it is impossible to see hummocks before stumbling over them, and walking is consequently very tiresome without skis, which more or less 'feel their way' along.

Distances in this milky light are very deceiving. A piece of dog dung a hundred yards away may look like a far-distant house. Because of the uniform white glare from the surface below, from which men's eyes are not naturally protected, sun blindness is likely unless protective goggles are worn. A very painful experience once it is incurred, something like burning hot sand in your eyes. We had no trouble sleeping safely out on the trail at temperatures around 50°F below zero. Our sleeping bags were caribou hide with the hair turned inward, with thin woollen blankets inside. When crawling out in the morning one had to exhaust the bag of its warm air, lest the vaporous moisture condense and freeze, making a very uncomfortable sleeping place for the next night.

For clothing, we departed somewhat from previous arctic experience and wore separate thin layers of wool (successive sweaters). These sweaters we put on or took off, according to temperature and the degree of exertion needed at the time. On the trail we carried an outer natural-fur parka, but did not usually wear it (even at fifty-below temperatures) until we made a long halt, or when outside in the evening, and so on.

In general, wind was more important than temperature as a factor in comfort. On one winter day in July, 1929, two other men and I started out on skis for a short local exploration. It was 9 A.M., and still completely dark, when we left camp. The temperature was colder than 70 below -actually - 72°F, as I remember it. Fortunately, however, there was no wind. It was completely calm; the sky was clear, with the intensely bright southernhemisphere stars (Canopus, the Southern Cross, and so on) swinging slowly around over our heads as the morning wore on, interspersed with the palegreen, flickering curtains of the aurora australis.

Starting off cool in thick clothing, we got well warmed up in a half-hour of skiing across the sand-like antarctic snow, which does not melt slightly under your skis like snow in warmer climates, providing a slippery surface. Hence skiing calls for heavy exertion in order to make any progress. Strange as it may seem, though at 9·30 A.M. the temperature was still 70 below and it was still dark, we were going forward barehanded and almost stripped to the waist. The only time we really felt the cold was on slight downslopes where we could slide along at six or seven miles an hour. Here the lack of heavy exercise, combined with the slight wind of our forward speed, made us immediately feel the bitterness of the Antarctic cold. We continued thus in the darkness until around noon, when the sun appeared for an hour or so as a faint glow on the northern horizon.

To find the day-to-day march of normal temperature in your own particular American region, you must consult a Weather Bureau publication entitled 'Normals of Daily Temperature of the United States.' But by averaging the normals for Washington City with those of certain inland cities in the same latitude, we can obtain something fairly representative of a large (central and eastern) portion of the country. Atlantic seaboard temperatures average two or three degrees Fahrenheit warmer in winter than inland temperatures for the same latitude; in summer the seacoast may average a degree or two cooler. Also, yearly maximums and minimums on the seacoast come several days later than corresponding maximums and minimums inland.

The northern United States (latitude of Boston) averages about five degrees Fahrenheit colder than this table. The southern United States (latitude of Savannah) averages about ten to fifteen degrees warmer than this table in winter and about three to five degrees warmer in summer. Temperatures in the Rockies and other high mountain regions vary widely with altitude and topography. Temperatures on the Pacific coast are largely conditioned by the water temperatures of the eastern Pacific Ocean, and by topography. In general,

Pacific seaboard temperatures show far less annual change than inland temperatures, and yearly maximums and minimums are delayed a month or so longer. Summer is typically cool and foggy, because of cold sea water that wells up from the great depths along the west coast.

The normal-temperature table printed above shows many well-known but interesting facts. The time of greatest cold comes around January 25, lagging more than a month behind the sun. The temperature remains almost constantly cold for a while; then, with the coming of spring, it begins to rise, slowly at first, later at the rate ½°F per day or 10° to 15°F per month. But the month of April is nearly as cold as the month of November; though many people, rushing out scantily clothed into the brightening sun, fail to realise the fact, and return indoors to meditate on it while nursing a spring cold. In the northernmost parts of the United States winter lingers into April, perhaps into May.

A period of nearly uniform summer heat comes in July and August, the maximum (around July 25) again lagging more than a month behind the sun, which has already turned back southward; and this summer heat is surprisingly uniform, by and large, over the whole United States. Then the temperatures begin to fall, slowly at first, then more rapidly, until by midautumn the rate of decrease is about equal to the rate of increase in spring. Finally the winter snows begin, and the snow itself, during the ensuing months, tends to hold air temperatures below freezing by reason of its reluctance to melt and its high reflecting power.

Tabulated observations at various levels show the following prevailing directions and velocities: altitude 1 mile, mostly westerly, averaging about 15 m.p.h. in summer and about 25 m.p.h. in winter; 2 miles, mostly westerly, averaging around 18 to 28 m.p.h.; 4 miles, mostly westerly, averaging about 25 to 40 m.p.h.; 6 miles, westerly to west-northwesterly, averaging about 40 m.p.h.

❐

7

Biodegradation of Waste

BIODEGRADATION OF WASTE

Biodegradation is nature's way of recycling wastes, or breaking down organic matter into nutrients that can be used by other organisms. "Degradation" means decay, and the "bio-" prefix means that the decay is carried out by a huge assortment of bacteria, fungi, insects, worms, and other organisms that eat dead material and recycle it into new forms.

In nature, there is no waste because everything gets recycled. The waste products from one organism become the food for others, providing nutrients and energy while breaking down the waste organic matter. Some organic materials will break down much faster than others, but all will eventually decay. By harnessing these natural forces of biodegradation, people can castes and clean up some types of environmental contaminants. Through composting, we accelerate natural biodegradation and convert organic wastes to a valuable resource.

Wastewater treatment also accelerates natural forces of biodegradation. In this case the purpose is to break down organic matter so that it will not cause pollution problems when the water is released into the environment. Through bioremediation, microorganisms are used to clean up oil spills and other types of organic pollution. Composting and bioremediation provide many possibililites for student research.

Wastewater Treatment

Wastewater treatment uses microbes to decompose organic matter in sewage. If too much untreated sewage or other organic matter is added to a lake or stream, dissolved oxygen levels will drop too low to support sensitive species of fish and other aquatic life. Wastewater treatment systems are designed to digest much of the organic matter before the wastewater is released so that this will not occur. Treatment systems use physical, chemical, and biological processes:

- Primary treatment physically removes large solids using grates, screens, and settling tanks.
- Secondary treatment promotes growth of bacteria and other microbes that break down the organic wastes. These biodegradation processes also take place in streams, lakes, and oceans, but the purification systems in nature can easily be overloaded with input of too much organic waste. Secondary treatment prevents this type of pollution by degrading most of the organic matter before the water is released into the environment.
- Tertiary treatment is used only where it is needed to protect the receiving waters from excess nutrients. In tertiary treatment, the concentrations of phosphorus or nitrogen are reduced through biological or chemical processes.
- Disinfection kills disease-causing organisms, most commonly through chlorination.

Sludge, the collection of solids that are removed during wastewater treatment, requires processing to reduce odour and water content. Depending on the disposal method, the sludge also may undergo treatment to decompose organic matter or kill disease-causing organisms.

Wastewater Treatment Principles

Sewage is the wastewater released by residences, businesses and industries in a community. It is 99.94 per cent water, with only 0.06 per cent of the wastewater dissolved and suspended solid material. The cloudiness of sewage is caused by suspended particles which in untreated sewage ranges from 100 to 350 mg/l. A measure of the strength of the wastewater is biochemical oxygen demand, or BOD5. The BOD5 measures the amount of oxygen microorganisms require in five days to break down sewage. Untreated sewage has a BOD5 ranging from 100 mg/l to 300 mg/l. Pathogens or disease-causing organisms are present in sewage. Coliform bacteria are used as an indicator of diseasecausing organisms. Sewage also contains nutrients (such as ammonia and phosphorus), minerals, and metals.

Ammonia can range from 12 to 50 mg/l and phosphorus can range from 6 to 20 mg/l in untreated sewage. Sewage treatment is a multi-stage process to renovate wastewater before it reenters a body of water, is applied to the land or is reused. The goal is to reduce or remove organic matter, solids, nutrients, disease-causing organisms and other pollutants from wastewater.

Each receiving body of water has limits to the amount of pollutants it can receive without degradation. Therefore, each sewage treatment plant must hold a permit listing the allowable levels of BOD5, suspended solids, coliform bacteria and other pollutants. The discharge permits are called NPDES permits which stands for the National Pollutant Discharge Elimination System.

Preliminary Treatment

Preliminary treatment to screen out, grind up, or separate debris is the first step in wastewater treatment. Sticks, rags, large food particles, sand, gravel, toys, etc., are removed at this stage to protect the pumping and other equipment in the treatment plant. Treatment equipment such as bar screens, comminutors (a large version of a garbage disposal), and grit chambers are used as the wastewater first enters a treatment plant. The collected debris is usually disposed of in a landfill.

Primary Treatment

Primary treatment is the second step in treatment and separates suspended solids and greases from wastewater. Waste-water is held in a quiet tank for several hours allowing the particles to settle to the bottom and the greases to float to the top. The solids drawn off the bottom and skimmed off the top receive further treatment as sludge. The clarified wastewater flows on to the next stage of wastewater treatment. Clarifiers and septic tanks are usually used to provide primary treatment.

Secondary Treatment

Secondary treatment is a biological treatment process to remove dissolved organic matter from wastewater. Sewage microorganisms are cultivated and added to the wastewater. The microorganisms absorb organic matter from sewage as their food supply. Three approaches are used to accomplish secondary treatment; fixed film, suspended film and lagoon systems.

Fixed Film Systems

Fixed film systems grow microorganisms on substrates such as rocks, sand or plastic. The wastewater is spread over the substrate, allowing the wastewater to flow past the film of microorganisms fixed to the substrate. As organic matter and nutrients are absorbed from the wastewater, the film of microorganisms grows and thickens. Trickling filters, rotating biological contactors, and sand filters are examples of fixed film systems.

Suspended Film Systems

Suspended film systems stir and suspend microorganisms in wastewater. As the microorganisms absorb organic matter and nutrients from the wastewater they grow in size and number. After the microorganisms have been suspended in the wastewater for several hours, they are settled out as a sludge. Some of the sludge is pumped back into the incoming wastewater to provide "seed" microorganisms. The remainder is wasted and sent on to a sludge treatment process. Activated sludge, extended aeration, oxidation ditch, and sequential batch reactor systems are all examples of suspended film systems.

Lagoon Systems

Lagoon systems are shallow basins which hold the waste-water for several months to allow for the natural degradation of sewage. These systems take advantage of natural aeration and microorganisms in the wastewater to renovate sewage.

Final Treatment

Final treatment focuses on removal of disease-causing organisms from wastewater. Treated wastewater can be disinfected by adding chlorine or by using ultraviolet light. High levels of chlorine may be harmful to aquatic life in receiving streams. Treatment systems often add a chlorine-neutralizing chemical to the treated wastewater before stream discharge.

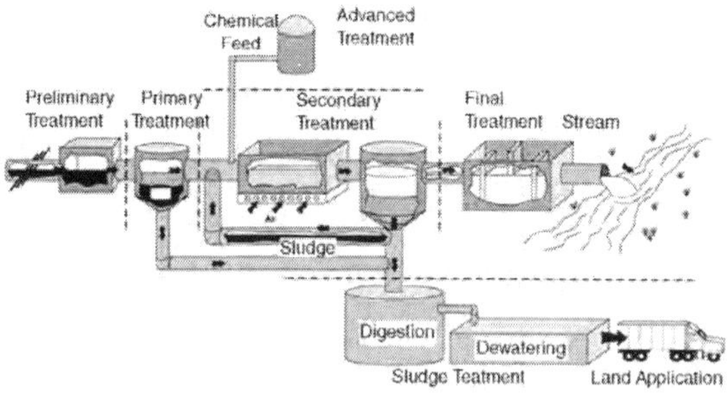

Fig. 7.1: Sludge Treatment

Advanced Treatment

Advanced treatment is necessary in some treatment systems to remove nutrients from wastewater. Chemicals are sometimes added during the treatment process to help settle out or strip out phosphorus or nitrogen. Some examples of nutrient removal systems include coagulant addition for phosphorus removal and air stripping for ammonia removal.

Sludges

Sludges are generated through the sewage treatment process. Primary sludges, material that settles out during primary treatment, often have a strong odour and require treatment prior to disposal. Secondary sludges are the extra microorganisms from the biological treatment processes. The goals of sludge treatment are to stabilize the sludge and reduce odours, remove some of the water and reduce volume, decompose some of the organic matter and reduce volume, kill disease causing organisms and disinfect the sludge. Untreated sludges are about 97 per cent water.

Settling the sludge and decanting off the separated liquid removes some of the water and reduces the sludge volume. Settling can result in a sludge with about 96 to 92 per cent water. More water can be removed from sludge by using sand drying beds, vacuum filters, filter presses, and centrifuges resulting in sludges with between 80 to 50 per cent water.

This dried sludge is called a sludge cake. Aerobic and anaerobic digestion are used to decompose organic matter to reduce volume. Digestion also stabilizes the sludge to reduce odours.

Caustic chemicals can be added to sludge or it may be heat treated to kill disease-causing organisms. Following treatment, liquid and cake sludges are usually spread on fields, returning organic matter and nutrients to the soil. Wastewater treatment processes require careful management to ensure the protection of the water body that receives the discharge. Trained and certified treatment plant operators measure and monitor the incoming sewage, the treatment process and the final effluent.

Wastewater Treatment Regulations

Clean water has been a concern nationwide Congress adopted the Clean Water Act to protect the waters of the nation. Through this act. Environmental Protection Agency and corresponding state agencies were given the responsibility to regulate activities that threaten the quality of the nation's water resources.

In the Federal Clean Water Act, Congress adopted a comprehensive water policy for the nation and set as a national goal the elimination of pollutant discharges to the navigable waters. An interim goal was set to insure that all navigable waters would be fishable and swimmable

To reach these goals:

- No one has the right to pollute the navigable waters of the United States. Dischargers are required to obtain permits.
- Permits shall set limits on the concentration of the pollutants being discharged. A violation of the limits carries a penalty of fines or imprisonment.
- The best technology available shall be used to control the discharge of pollutants.

Each state, through a designated regulatory agency, issues discharge permits and enforces the discharge limits. The Ohio Water Quality Standards specify that all surface waters of the state must be free from the following pollutants as a result of human activity: suspended solids, floating debris, colour, odour, toxic substances and nutrients that create nuisance growths of weeds and algae.

Permits for wastewater treatment systems serving single family homes, two- or three-family dwellings are issued by local health departments. Permits for all wastewater treatment systems serving more than 2 homes or buildings larger than a 3-family dwelling are issued by the Ohio Environmental Protection Agency.

Two different permits are issued by the Ohio Environmental Protection Agency. All systems must first obtain a Permit to Install (PTI) before a treatment plant can be constructed. The Permit to Install is issued after the plans for construction, operation and management are examined to insure that the receiving stream will be protected.

If the discharge from the treatment plant enters a stream, a NPDES permit is required. The NPDES permit specifies the maximum allowable level of total suspended solids, biochemical oxygen demand, nutrients and bacteria that can be discharged to a stream as well as the minimum level of dissolved oxygen that must be present in the discharge. The levels specified in the NPDES permit are determined by the condition of the receiving stream. Therefore, NPDES permits are subject to change every 5 years as water quality concerns change throughout Ohio. Discharging raw sewage from a dwelling is considered a public nuisance. For systems serving a 1-, 2- or 3-family dwelling, the local health department is the responsible regulatory agency. The penalty for creating a public nuisance is a third degree misdemeanor which is punishable by not more than 60 days in jail and/or a $500 fine. In addition, the violator may be required to remove the sewage from public or private property or the waters of the state.

Discharges of pollutants from a multiple dwelling is under the jurisdiction of the Ohio Environmental Protection Agency. Discharges may be liable for civil penalties of up to $10,000 for each day of violation. In addition, criminal penalties can be assessed or up to one year of imprisonment or both. A "Connection Ban" can also be issued by the Ohio Environmental Protection Agency which prohibits the construction or installation of home sewage disposal systems when it is shown that pollutants from new homes will be discharged into the waters of the state.

The Ohio Environmental Protection Agency issued 25 administrative orders totaling $80,000 in fines to Ohio communities for wastewater treatment violations.

This publication may contain pesticide recommendations that are subject to change at any time. These recommendations are provided only as a guide. It is always the pesticide applicator's responsibility, by law, to read and follow all current label directions for the specific pesticide being used. Due to constantly changing labels and product registrations, some of the recommendations given in this writing may no longer be legal by the time you read them. If any information in these recommendations disagrees with the label, the recommendation must be disregarded.

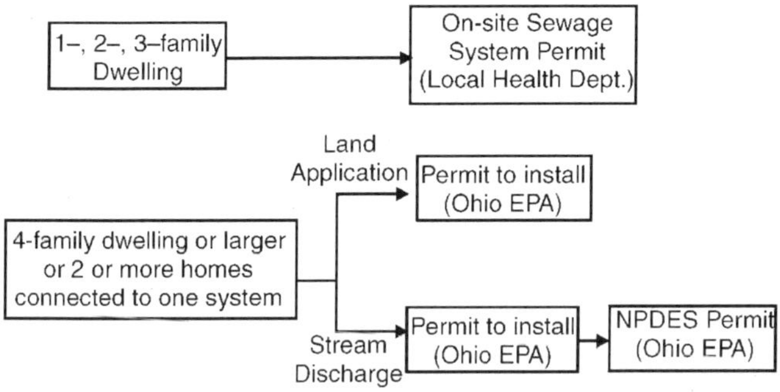

Composting

Composting refers to the process of combining organic materials such as weeds, grass clippings, and food scraps under conditions that enhance the rate at which they decompose. In the United States, organic wastes make up 20% to 40% of what we throw away.

Composting provides an opportunity to keep these organic materials out of landfills and incinerators, using them instead to create a useful end product. Of course, composting isn't as easy as simply throwing organic waste into big piles and waiting for it to decompose. If moist food scraps or grass clippings are placed in a container and left to sit for a week or two, the result is likely to be a stinky mess that attracts flies.

Given the proper conditions, these same organic materials can be composted to produce a material that looks and smells like rich soil and is prized by gardeners. Clearly the second option is preferable, and the key question is how to provide the proper conditions for problem-free composting. The answer lies in paying attention to moisture content, airflow, and the mixture between materials that are high in carbon and those that are high in nitrogen.

Composting takes place at a wide range of scales, from huge commercial or municipal operations to simple backyard bins or heaps. It can also be carried out indoors in containers such as garbage can bioreactors or worm bins. For research purposes, composting can even take place in soda bottle bioreactors, which are small and inexpensive enough to enable students to build their own individualized systems. Once made, compost can be used for gardening projects or for plant growth experiments ranging from nutrient analysis of compostenriched soils to use of composts to suppress plant diseases

Compost Ingredients

Garden supply stores and catalogs often sell compost "starters," which supposedly speed up the composting process. Develop a recipe for a compost starter and design a research project to test its effect on the compost temperature profile.

How well do human nutrition concepts apply to compost microorganisms? The microbes get a "sugar high," demonstrated by a quick, high temperature peak when fed sugary foods, compared with a longer but lower peak for more complex carbohydrates.

Measure the pH of a number of different compost mixes. How does the pH of initial ingredients affect the pH of finished compost?

Some instructions call for adding lime to increase the pH when compost ingredients are mixed. Other instructions caution to avoid this because it causes a loss of nitrogen. How does adding various amounts of lime to the initial ingredients affect the pH of finished compost.

Microorganisms

Composting recipes sometimes call for inoculating the pile by mixing in a few handfuls of finished compost. Is there any observable difference in appearance of microbes between systems that have and have not been inoculated.

The pH of the initial compost ingredients affect the populations of microorganisms during composting.

Worm Composting

Do organic wastes in compost break down more readily in the presence of worms than through composting that depends solely on microbial decomposition.

In some experiments, plants have not show increased growth when planted in fresh worm castings. Does aging or "curing" worm castings increase their ability to enhance plant growth? Are there chemical differences between fresh and older worm castings? Should worm compost be mixed with soil before being used to grow plants.

Different food sources affect reproductive and growth rates of red worms (Eisenia fetida). Red worms grow best in wastes with pH between 5.0 and 8.0.

Effects of Compost on Plant Growth

Some leaves, such as those of black walnut or eucalyptus trees, contain chemicals that inhibit growth of other plants. Are these compounds broken down by composting. Finished compost is near neutral pH. Water in which compost has been soaked (often called compost tea) is said to be beneficial to plants. Can you design experiments to test whether different types, concentrations, and amounts of compost tea enhance plant growth? In China, farmers dig parallel trenches and fill them with organic wastes mixed with cocoons of Eisenia fetida. Soybeans planted in rows between the trenches are highly productive.

Building a Two-Can Bioreactor

Purpose

Two-can bioreactors are designed to be used as small-scall indoor composting units for families, and for composting as an educational tool in the classroom.

Materials

- 32-gallon plastic garbage can
- 20-gallon plastic garbage can
- Drill
- Brick
- Spigot (optional)
- Duct tape (optional)
- Insulation (optional)

Construction

- Using a drill, make 15 to 20 holes (0.5″ to 1″ diameter) through the bottom of the 20-gallon can. Next drill three rows of holes through the sides of this can, six to eight inches apart with four to five inches between rows, ending about two inches below where the can expands at the top.
- Place a brick or some other object in the bottom of the 32-gallon can. This is to separate the leachate from the compost and allow for its measurement and addition back into the compost pile. The leachate, often referred to as "compost tea," is rich in nutrients which may be in a form readily usable by plants. If not used right away on growing plants, pour the leachate back into the compost. Excessive leachate can be responsible for foul smells. If your sytem produces enough leachage to cause odour problems, your initialscompost mixture was probably too wet.
- Variations on the design: — Add insulation to the barrels (inner and outer) with duct tape. — Include a spigot to draw off the leachate. — Add a layer of old compost, wood chips, or soil inside the outer barrel. This will allow the leachate to be absorbed and may cause fewer leachate/odour issues. A system of 10-gallon plastic garbage cans that can fit inside 20-gallon cans can be substituted if space is a problem. The smaller system may operate at lower temperatures. This should not affect the final product; it will just take longer before the product can be used.

Bioremediation

"Remediate" means to solve a problem, and "bio-remediate" means to use biological organisms to solve an environmental problem such as contaminated soil or groundwater.

In a non-polluted environment, bacteria, fungi, protists, and other microorganisms are constantly at work breaking down organic matter. What would occur if an organic pollutant such as oil contaminated this environment? Some of the microorganisms would die, while others capable of eating the organic pollution would survive.

Bioremediation works by providing these pollution-eating organisms with fertilizer, oxygen, and other conditions that encourage their rapid growth. These organisms would then be able to break down the organic pollutant at a correspondingly faster rate. In fact, bioremediation is often used to help clean up oil spills.

Bioremediation of a contaminated site typically works in one of two ways. In the case described above, ways are found to enhance the growth of whatever pollution-eating microbes might already be living at the contaminated site. In the second, less common case, specialized microbes are added to degrade the contaminants. Bioremediation provides a good cleanup strategy for some types of pollution, but as you might expect, it will not work for all. Bioremediation may not provide a feasible strategy at sites with high concentrations of chemicals that are toxic to most microorganisms. These chemicals include metals such as cadmium or lead, and salts such as sodium chloride.

Nonetheless, bioremediation provides a technique for cleaning up pollution by enhancing the same biodegradation processes that occur in nature.

Depending on the site and its contaminants, bioremediation may be safer and less expensive than alternative solutions such as incineration or landfilling of the contaminated materials. It also has the advantage of treating the contamination in place so that large quantities of soil, sediment or water do not have to be dug up or pumped out of the ground for treatment.

Bioremediation is a process that can destroy pollutants by microbial action. Certain bacteria can use nutrients and oxygen to degrade volatile organic compounds and toxic compounds in soils. In the 1980s, bans on disposal of hazardous waste at landfills led to the increased primary use of disposal alternatives such as bioremediation.

However, little information has been available on the airborne emissions associated with these processes. This project was designed to help determine whether airborne emissions from bioremediation operations could pose an air pollution problem.

Decay and Renewal

Developed by the Cornell Environmental Inquiry Progammeme, Decay and Renewal consists of a student edition and teacher's guide designed to enable high school students to carry out authentic research. By using scientific protocols to investigate the processes in wastewater treatment, composting, landfilling, and bioremediation of contaminated sites, Decay and Renewal teaches important science concepts within the context of environmental issues.

Decay and Renewal includes:

- An overview of basic principles of biodegradation, the assortment of biological processes that cause organic matter to decay,
- Simple but authentic research protocols that measure biodegration in compost, landfills, wastewater treatment facilities, bioremediation sites, and streams,
- A progression of worksheets that guides students through each step of research, providing structure but flexibility in designing and conducting meaningful experiments, and
- Guidelines for integrating peer review and other collabourative knowledge-building into classroom science

Bacterial Biodegradation of High Molecular Weight Polycyclic Aromatic Hydrocarbons

Polycyclic aromatic hydrocarbons (PAH) are among the most common pollutants found at contaminated industrial sites, and in fact around the world.

Although PAH are naturally occurring compounds and essentially ubiquitous at low concentrations in the terrestrial environment, high concentrations of PAH are found in contaminated soils at wood treating facilities, at sites formerly used to produce manufactured gas, and at petroleum processing operations. There are a wide range of PAH, from the simple tworing compound naphthalene to large multi-ring compounds. We arbitrarily define those PAH with two or three rings as low molecular weight (LMW) species, and those with four or more rings as high molecular weight (HMW) species. One of the characteristic features of PAH contamination is that there is always a complex mixture of PAH, often in conjunction with a variety of other hazardous chemicals.

Environmental Protection Agency currently regulates 16 PAH compounds as priority pollutants in water, and generally considers these same compounds as "total PAH" (tPAH) in contaminated soils. The 16 regulated PAH comprise both low and high molecular weight species, and seven of them are designated as known human carcinogens. All of the carcinogenic PAH (cPAH) are high molecular weight compounds.

In addition to their toxicity, PAH as a class are extremely hydrophobic chemicals. Naphthalene is among the most water-soluble of the PAH, and its solubility in water is only about 30 mg/L. The situation becomes much worse with increasing molecular weight: chrysene (a four-ring PAH) and benzo[a]pyrene (a five-ring compound) are soluble in water in the low part per billion (mg/L) range. By comparison benzene, normally considered to be a water-immiscible chemical, has an aqueous solubility of about 2,000 mg/L. Despite the very low solubility of PAH in water, PAH-contaminated materials have been shown to be toxic in a variety of bioassays.

Microbiologists have known for a long time that microorganisms can act on most of the PAH of concern. For this reason, PAH-contaminated soils and sediments are generally considered to be candidates for bioremediation. In many of the studies in which PAH degradation in contaminated soil or sediment has been studied, however, the higher molecular weight compounds have not been removed completely. We generally do not know whether this apparent recalcitrance is due to inherently greater difficulty in metabolism of these compounds by microorganisms, a lack of bioavailability caused by their low solubility in water, or both. It is also not usually known whether microbial activity leads to complete mineralization (conversion to carbon dioxide and water) of individual PAH or only partial transformation to other organic products. Some fungi, will metabolize PAH only partially in pathways that seem to be analogous to mammalian detoxification mechanisms. While fungi do not appear to use PAH as primary carbon sources for growth, bacteria have been known for many years to be able to use several of the low molecular weight PAH as growth substrates; in fact, some of the groundbreaking work on aerobic bacterial metabolism of aromatic hydrocarbons included studies on naphthalene metabolism.

Ironically the depth of study on bacterial metabolism of naphthalene and, to a lesser extent, phenanthrene has been accompanied by only sporadic study of the metabolism of other PAH. Until the late 1980s, it was generally believed that PAH with four or more rings could not serve as growth substrates for bacteria, and could only be degraded at the expense of other growth substrates.

Since then a handful of reports have appeared in the literature describing bacterial growth on the four-ring compounds fluoranthene, pyrene, and chrysene, but there is still no evidence of bacterial growth on PAH with more than four rings. There are also several studies in which the high molecular weight PAH were observed to be mineralized in the absence of growth, either by pure cultures or in complex environmental samples such as soil or sediment. We still know relatively little, however, about the mechanisms by which the high molecular PAH are degraded by bacteria.

Our work on bacterial degradation of the high molecular weight PAH began by isolating bacteria from PAH-contaminated soils. In all, we obtained ten isolates from eight different sites and one more PAH-degrading strain from another investigator. We have used these bacteria to study various aspects of PAH biodegradation, which focus on two key issues: whether the high molecular weight PAH are degraded by the same metabolic pathways a given bacterium uses to degrade the low molecular weight PAH; and how the low aqueous solubility of these compounds influences their availability to bacteria, which is prerequisite to degradation.

Mineralization of High Molecular Weight PAH

All of the bacteria we isolated from PAH-contaminated materials were obtained by enrichment culture techniques using phenanthrene as the sole carbon source. This clearly biases our observations to only those bacteria which can grow on phenanthrene, but we had hypothesized that such bacteria might be able to degrade a range of other PAH. Our hypothesis was based on the knowledge that these organisms are always exposed to a mixture of PAH, all of which are, to varying extents, structural analogues.

Over evolutionary time scales such conditions would be expected to select for microorganisms with versatile metabolic capabilities. We also knew that organisms able to degrade but not grow on the higher molecular weight compounds needed to grow on something, so the low molecular weight PAH seemed like the best choice with which to begin.

We have studied each of the 11 bacterial isolates for their ability to grow on or degrade 13 PAH, ranging from two- to five-ring compounds. Most of the bacteria have a broad range of PAH substrates they can metabolize. Of most interest in terms of maximizing the biodegradation of these compounds, however, is mineralization. So far we have studied the ability of nine of the strains to mineralize pyrene, chrysene, benzanthracene and benzopyrene (the latter three compounds are all cPAH).

First, none of the organisms was able to mineralize pyrene under the conditions used in this experiment. Second, every organism either mineralized all three of the remaining high molecular weight substrates or did not mineralize any of them. These findings suggest that pyrene metabolism may not be linked directly to the metabolism of the other PAH, and that there may be a relationship among the other three PAH.

Induction of High Molecular Weight PAH Degradation

Many heterotrophic bacteria have a broad range of organic compounds they can grow on or degrade. To conserve energy for vital growth-related processes, the enzymes needed by these bacteria to degrade a particular compound are not normally synthesized unless the compound is present in the medium. Thus, biodegradation pathways in bacteria are normally induced, or "turned on," by the presence of either the parent compound or one of the pathway intermediates. If we are to learn how to control the biodegradation of compounds such as high molecular weight PAH, we clearly must understand how the relevant enzymes are turned on. This is particularly important for those compounds that do not serve as growth substrates for bacteria, since they are not likely to be able to induce their own degradation.

The only well studied example of the induction of PAH degradation in bacteria is that of naphthalene degradation. The complete pathway for conversion of naphthalene to simple, common metabolic end products is inducible by the intermediate salicylate in the bacteria for which it has been studied. In other words, the enzymes required for naphthalene metabolism are elevated to high levels in naphthalene-degrading bacteria when salicylate is introduced into the medium. (The astute reader might ask how salicylate is formed when naphthalene is the only carbon source in the medium; the answer is that in such cases the organism produces very low, or constitutive, levels of the required enzymes at all times, which then metabolize the naphthalene to form salicylate).

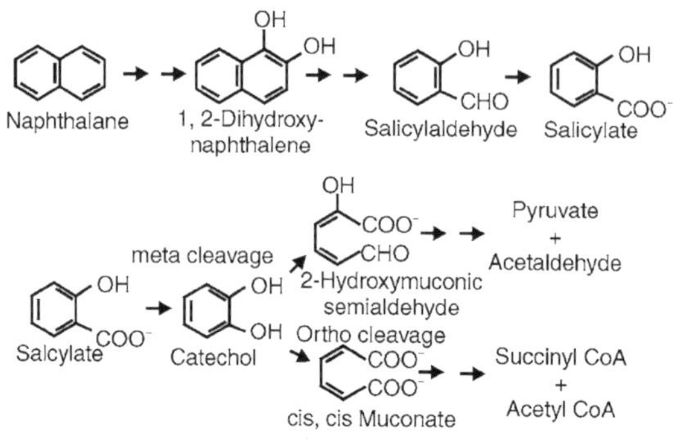

Fig. 7.2: The Principal Pathway for Acrobic Metabolism of Napthalene by Pseudomonads.

We have focused our attention on the inducibility of HMW PAH degradation in one of the organisms in our collection, Pseudomonas saccharophila P15. This organism was isolated several years ago by Will Stringfellow, who also conducted in-depth studies on its physiology. Both phenanthrene and salicylate induce the mineralization of benzanthracene, chrysene, and benzopyrene by P. saccharophila.

None of the three compounds are used as growth substrates by the organism, so these results suggest that the metabolism of high molecular weight PAH in P. saccharophila P15 is linked to the metabolism of a low molecular weight compound. We do not yet know, however, if the different PAH are metabolized by a common pathway or if different pathways are regulated through a common mechanism.

We hope to understand eventually if the metabolism of high molecular weight PAH by the other organisms in our collection is inducible by salicylate or other inducers of low molecular weight PAH metabolism. If we find such inducibility to be a general phenomenon, then we may have a tool with which to aid the removal of HMW PAH in bioremediation projects.

Bioavailability

Those who study the biodegradation of poorly soluble organic compounds in soil or sediment usually assume that only the amount of a chemical that is present in the aqueous phase is available to the microorganisms capable of degrading that chemical. Some microbes may have their own mechanisms of improving the bioavailability of these compounds, however. Many of the bacteria that degrade aliphatic hydrocarbons are known to produce surfactants (so-called "biosurfactants") which can solubilize or emulsify the hydrocarbons, presumably making them available for degradation at a much faster rate.

Nobody knows whether there is a similar involvement of biosurfactants in the natural biodegradation of PAH. We screened our collection of bacteria for their ability to reduce the surface tension in the culture medium (an indicator of surfactant production) during growth on phenanthrene or naphthalene, but only one organism was able to do so significantly. Thus, we did not see evidence that biosurfactant production is a common feature of PAH metabolism by these bacteria. We were, however, recently able to demonstrate how synthetic surfactants can improve the rate of phenanthrene biodegradation in a well-defined experimental system

Another mechanism by which bacteria might improve their access to poorly soluble chemicals in soil or sediment is through their motility. Motility is a very common feature of soil bacteria, yet its role in biodegradation of pollutants remains virtually unexplored.

Chemotaxis is a special form of motility in which an organism swims towards chemoattractants or away from chemorepellents, both cases requiring the existence of concentration gradients of the attractant or repellent for a chemotactic response to occur. The reliance on concentration gradients seems to make chemotaxis an ideal mechanism by which bacteria might be able to locate sources of "food," which in the case of hydrophobic chemicals is likely to be distributed heterogeneously in the porous medium.

We have become interested in the potential effects of motility and chemotaxis on PAH biodegradation. The lack of previous work on environmentally relevant situations has left a wide scientific gap which needs to be filled, to understand both the importance of such mechanisms in in situ biodegradation reactions and whether such mechanisms can eventually be manipulated to our advantage. If, motility and/or chemotaxis are found to be important aspects of PAH (or other pollutant) biodegradation by soil bacteria, then we need to begin incorporating such phenomena into models that purport to simulate biodegradation processes in the subsurface.

Most of the bacteria identified are motile, and we screened most of them for chemotaxis to naphthalene, phenanthrene and known metabolites of these low molecular weight PAH. Only one of them, Pseudomonas stutzeri P16, seemed to have a chemotactic response to the chemicals of interest. Interestingly, strain appears to be chemotactic to 1-hydroxy-2-naphthoate and to salicylate, both metabolites that have been found extracellularly in media containing active PAH degraders.

However, we are learning that some of the methods traditionally used to study chemotaxis are not well suited to the study of chemotaxis towards compounds of low aqueous solubility.

To test our methods with a known chemotactic organism, we recently obtained the strain Pseudomonas putida G7 from Professor Caroline. We learned that the chemotactic response of strain G7 in a standard quantitative assay becomes increasingly detectable as the starting concentration of bacterial cells is decreased.

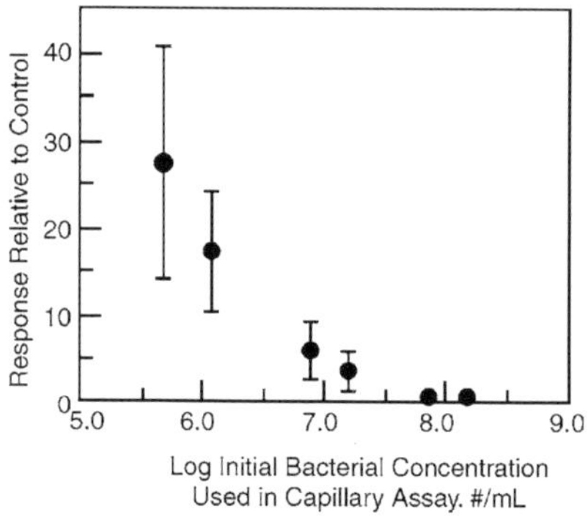

Fig. 7.3: Effect of initial Bacterial Concentration

The optimum response we observed was at a cell concentration nearly two orders of magnitude lower than the concentration typically used in the assay, which we attribute to the rapid loss of the naphthalene concentration gradient at high cell concentrations and the necessarily low starting concentration of naphthalene. Overall the response of strain G7 to naphthalene was variable yet strong. Consequently, we will focus on this organism in our subsequent efforts to understand the relevance of chemotaxis to PAH biodegradation. There is much to be learned about the bacterial biodegradation of polycyclic aromatic hydrocarbons, and we of course can only begin to scratch the surface. In another project, we are studying whether the concepts we have explored relative to the inducibility of PAH metabolism and the effects of surfactants on PAH bioavailability can be applied to the bioremediation of PAH-contaminated soils. It is always difficult to bridge the gap between basic and applied research, but we ultimately hope that each will provide insights that will shape the future direction of our own work and that of others, both in the research community and in practice.

Pollution Prevention in Resource-related Setting

Preventing pollution in a sensitive resource-related setting means thinking through all of the activities and services associated with the facility and planning them in a way that generates less waste. Waste prevention leads to thinking about materials in terms of reduce, reuse, recycle. The best way to prevent pollution is not to use materials that become waste problems. When such materials must be used, they should be reused onsite. Materials that cannot be directly reused should be recycled. Everyone associated with the facility must change their habits and adopt a more responsible attitude toward waste. This includes the ownership and management of the facility-the architects, engineers, designers, and contractors; the employees; and the visitors. Each of these groups needs to consider the issues so that no waste will be generated that adversely affects the environment.

Visible, Participatory Systems

The "out of sight, out of mind" mentality regarding waste is perpetuated because the systems that deal with waste problems are all behind the scenes and off-limits. Prevention can only be accomplished by paying attention to the issue up front, rather than waiting until after the problem has been created. An environmentally sound tourism facility would ensure the visibility of systems to prevent the generation of waste. Such systems require conscious participation by visitors, users, and operators. They should not dominate the experience of a visitor at the facility. If each visitor does his or her share, the facility can be operated in a more environmentally sound manner. An additional benefit is that this can lead to long-term changes in behaviour benefiting the participant and the earth.

Training and Maintenance

Waste prevention requires training the operators, educating all users of the system, and performing diligent maintenance. Most waste problems have been created because attention has not been paid. Conventional disposal systems are designed for ease of use with the user having little idea of how the systems operate or where the waste goes. Because waste prevention represents a change in the way activities are carried out, it requires an extra effort to ensure that these practices are maintained until they become routine. In situations with high turnover of both employees and visitors, continuous training and education would be essential.

Host and Visitor Attitudes

Preventing waste means taking a measure of responsibility for activities. This can conflict with the attitude that going on vacation means escaping responsibilities. Without an attempt to change this attitude, the necessary participation may be difficult to obtain.

Fostering a sense of responsibility in visitors is an important element of environmentally sound development in many other ways as well. One way to encourage this attitude is to convey to visitors that they are privileged visitors, and as a result, they may naturally feel obliged to follow certain guidelines established by their hosts. Facility operators should behave as hosts that care about making waste prevention work. Employees of the facility must care about it, and that caring must be visible to the visitors.

Garbage/Solid Waste Prevention Strategies

In planning for facilities, a comprehensive design strategy is needed for preventing generation of solid waste. A good garbage prevention strategy would require that everything brought into a facility be recycled for reuse or recycled back into the environment through biodegradation. This would mean a greater reliance on natural materials or products that are compatible with the environment.

Any resource-related development is going to have two basic sources of solid waste-materials purchased and used by the facility and those brought into the facility by visitors. The following waste prevention strategies apply to both, although different approaches will be needed for implementation:

- Use products that minimize waste and are nontoxic
- Compost or anaerobically digest biodegradable wastes
- Reuse materials onsite or collect suitable materials for offsite recycling

Use of Products that Minimize Waste

Much of the growing volume of garbage is from the use of disposable consumer products and excess packaging. Consideration must be given to materials or products that minimize waste disposal needs. Purchase items with minimal packaging; buy in bulk; replace disposable products with durable, reusable items. Durable, reusable products can be substituted for disposable ones, such as the use of returnable glass bottles instead of aluminum cans or the use of rechargeable batteries. Use of plastics for packaging is increasing, thereby replacing recyclable products and materials. Plastics, which account for about 20% of solid waste by volume, are nonbiodegradable, difficult or impossible to recycle, have a high volume to weight ratio, and are toxic when burned.

When selecting materials and goods, nothing should be purchased that will ultimately become toxic. Nontoxic materials can often be substituted for products that cause contamination problems during disposal.

Materials should be purchased locally whenever possible. Locally produced goods needing less transport and less storage should have less waste packaging. No matter how diligent people are in purchasing materials, certain items will need to be handled. There are two methods to considerbiodegradation or recycling. Technology and economics are continually changing and the system selected must have the flexibility to adjust to market conditions. The scope of composting or anaerobic digestion of organic materials and the availability of markets for recycled materials should be considered in planning for a facility. These decisions are site-specific and must be made at the outset before final design is completed. Factors such as land availability, local markets, and isolation all contribute to that decision.

Biodegradation

In the process of biodegradation, microorganisms break down the products of other living things and incorporate them back into the ecosystem. Biodegradable or bioconvertible material includes anything that is organic. Plastics are not considered includable, despite industry contention that they are. Most of the organic components of garbage, such as paper and food wastes, can be eliminated through composting. Between 60% and 75% of the solid waste is bioconvertible.

In a tourism development, the greatest fraction of waste would be generated by the support and maintenance services. Wastes that are bioconvertible include newspapers, magazines, and wet wastepaper in the kitchen and restaurants. Also included are food waste, cardboard materials, paper office supplies, and leaves, grass, and tree trimmings. The solid fraction of the sewage waste is also considered available for the bioconversion process, and is often the most costly to dispose of otherwise, usually in a special landfill.

The biodegradable or bioconvertible percentage of the waste stream is large enough to consider at least two options for the conversion process. Two obvious options for conversion are composting and anaerobic digestion.

Composting

Composting is a familiar concept, and is used for handling yard waste and even sewage sludge. Both of these organic wastes require mixing of other materials to achieve a nutrient balance. Large chunks of relatively inert material (most commonly wood chips) add bulk and aeration to make the process work. This is typically done in open windrows or piles, with mixing done daily to provide aeration and homogeneity. This takes land space on a drainable surface, and a collection of any runoff for redispersal of the liquid to the process. It produces a quality soil amendment and reduces the bulk of the original material by approximately 40-50%. Composting does off-gas ammonia and carbon dioxide and produces offensive orders.

It typically takes 50-60 days to process. Screening of the final product is necessary to remove the bulking material, and provide granulation before use in the soil bed. The use of this product as a soil amendment is valuable, particularly in the tropics, because the soil is essentially sterile, with only about 2% organic content. Affected by humidity, rainfall, temperature, and normal soil activity, the organic material placed on the soil will typically last only 30- 40 days in the tropics. In a temperate climate, that same material may last as long as six months.

Most composting operations of the size typical for communities could be suitable for tourism development use, considering the opportunity to also process construction debris and landscape waste. This will typically be the smallest scale. The construction debris process equipment can be transferred directly to the operations management and be capable of handling the full volume of waste.

Anaerobic Digestion

Anaerobic digestion is used extensively worldwide for the processing of food waste, animal waste, the solids of human waste systems, and for the total array of solid waste such as waste paper, green waste, and landscape waste. This wet fermentation process converts the waste stream into three usable by-products:

- Biogas-an energy-rich gas stream, comparable to natural gas, that can be used to offset the cost of energy utilities of the development
- A high quality organic fertilizer solid that may be useful in landscaping efforts or even crop production;
- A dilute liquid organic fertilizer that may be used in drip irrigation as an additive to any planting progammeme, for foliar feeding of ornamentals, or in landscape plots for replenishing native or endangered plant species of plants.

The anaerobic digestion process does need management supervision in startup. Design of the system should be simple and modular as well as easy to operate. The modular design allows seasonal loading rather than using one large vessel. This also provides for stages of digestion that help the natural biology of the fermentation reach its optimum capacity. While this requires more capital investment, the savings to be made in maintaining operations through seasonal fluctuation will prove to be cost effective. The advantages of this system are in the smaller space it would require and in the energy it would produce, which may make the facility more self-sufficient. The value of the by-products cannot be understated, particularly in the case of remote development.

Recycling

A material doesn't become waste until it is thrown in the garbage can. If a material can be reused it is a resource, not waste. Reuse is the best form of recycling. There are markets in the United States, Europe, Asia and other parts of the world for many recyclable materials including aluminum, paper, glass, steel, and some types of plastics. These may be available to a facility depending on location and the quantity of the materials generated. Recycling can be maximized through the purchase of products for which there is a ready market as recycled materials. The feasibility of recycling a given material depends on the price offered by a buyer for that material and shipping costs, both of which may vary over time. However, if no other reuse or recycling option exists, materials used at the facility should be recycled even at a net loss rather than sent for disposal. In circumstances where there is no available market for a given material, often a beneficial end use can be developed locally. Every effort should be made to work with the local community to determine if any of the materials generated by the facility can be used.-e.g., glass beverage containers can be ground up and incorporated into materials for construction and road building. Efficient recycling requires sorting of materials; convenient bins should be provided at the facility for the materials being recycled.

Offsite Disposal

If a garbage prevention strategy has been fully executed, actual remaining waste should be minimal. Remaining residuals mean that the facility is not entirely environmentally sustainable. All residuals must be collected separately and disposed of offsite. Although disposal of residual waste anywhere would have an adverse impact on the environment, in an environmentally sound development special care must be taken that this will not be borne locally. In most cases residuals must be returned to their place of origin. Toxic material residuals must be segregated and disposed of separately.

Materials Brought in and/or Purchased Onsite

Visitors and guests normally bring various materials that are consumed onsite, often leaving behind disposal problems. Products purchased onsite and in the surrounding community can exacerbate the problem. Such materials may include toxic and difficult to dispose of items (e.g., batteries, flash bulbs, instant film refuse, and repellents. Education, including what is incorporated in the written materials about the facility and what is available, is important in minimizing these types of waste materials.

DOMESTIC WASTE MANAGEMENT STRATEGIES

General Considerations

Domestic wastes in park and tourism developments come from toilets and urinals, showers, bathroom sinks, kitchen sinks, laundry facilities, and most floor drains. Varying amounts of water are used to carry these wastes through pipes to a treatment point. Different processing techniques are used to convert the wastes for reuse or disposal. True sustainable development would not permit direct disposal of either liquids or solids before reuse.

Responsible waste management recognizes the value of reducing water needs. Properly treated wastewater can be used for toilet flushing (after approved disinfection) and irrigation (landscape needs or agricultural plantings).

The solids separated from domestic waste may be incorporated into the solid wastes (garbage) and composted into an excellent soil amendment product or anaerobically digested to produce a gaseous energy source and a residue that can also be used as a soil amendment.

If wastewater recycling for toilet flushing is used without concurrent irrigation of vegetation, there will be a volume of excess liquid that must be sensitively disposed of.

This will be the amount of liquid wastes coming in each day from all sources, less the recirculated amount used for toilet flushing. Best disposal practice would use a suitable area for subsurface movement of the liquid through soils that provide good filtering and additional treatment before reaching any body of water. Direct discharge to a waterway should be the lowest priority of alternatives investigated.

Using treated, recycled wastewater for toilet flushing, rather than an equal volume of drinking-quality water, would save a large amount of water. There are also strong economic and environmental reasons for using one of more than 20 different low-volume flush toilets currently on the market, rather than the more conventional water-wasting models. In seawater development sites, toilets may be flushed with seawater and processed with septic tank and subsurface treatment and disposal. This provides a reliable system if the corrosive nature of seawater is addressed in selection of materials used in the piping, plumbing, pumping, and treatment systems.

Similar water conservation can be realised through the use of flowrestricted, fine-spray shower heads, and sink faucet aerators. Spring-loaded handles should be installed on faucets. Laundry waste volume can be reduced through the use of machines with suds-saver cycles.

Park and tourism developments that make use of water conservation devices should make an effort to educate their visitors on the amount of water (or energy) they have conserved during their stay, compared with the conventional development. It is important to monitor the conservation devices in order to evaluate their performance.

Specific Waste Treatment Technologies

Conditions at sustainable developments favour simple, reliable, passive techniques for waste treatment. Systems that have few moving parts, controls, and monitoring requirements should be used. Continuity of operation and maintenance knowledge is essential. Waste treatment systems must also be able to withstand seasonal fluctuations in use. Preparation for an idle period should be simple, and no energy should be required when the system is not being used. If an innovative technology is to be employed for waste treatment, it should have a conventional backup system, or be sufficiently redundant that multiple failures are unlikely.

Dry Toilets

A composting toilet consists of a large tank located directly below the toilet room. Wastes enter the tank through a large diameter chute connecting to the toilet, and decompose in an oxygen rich environment. No water is used for the toilet, but a bulking agent (such as wood shavings) is added to improve liquid drainage and aeration, and to provide fuel. A small fan draws air through the tank and up the vent pipe to ensure adequate oxygen for decomposition and odourless operation. Internal components (such as ducts, baffles, and rotating tines) enhance the composting process. The finished compost can be removed from the lower end of the tank about once each year. It can be used as a fertilized for soil.

Composting toilets need a mild temperature, moisture, fuel, and air to function. Liquid may have to be added to the tank to keep the compost pile moist during periods of little use or a bulking agent added periodically to improve the compost texture.

Composting toilets have several advantages over other systems-e.g., no water is used and only a small amount of energy is needed for an exhaust fan; valuable nutrients are used to benefit soils; proper maintenance requires little time.

There are some disadvantages to this system. The user is close to the sewage treatment systems and this bothers some visitors. Without proper maintenance, the tank can become anaerobic and unpleasant odours arise. Undesirable pests can take up residence in the composting tank. The availability of a bulking agent in some areas can also be a problem.

Anaerobic Waste Treatment

Anaerobic waste treatment (septic systems) is accomplished through microorganisms living in the wastewater. Anaerobic microorganisms work in an environment where there is no free oxygen. Complex reactions and interactions take place with the resulting generation of some offensive gaseous by-products. These unpleasant odours are actually an indication of an efficient progression of the anaerobic process to remove the pollutants from the waste stream.

This type of system is reasonable to consider for smaller developments, where the by-product gases can be separated from occupancy areas, dissipated with good air movement, or neutralized with soil or carbon filters. Since slow treatment means longer holding periods (shallow depth tanks), large, isolated treatment and disposal areas are needed. Treated wastewater (effluent) is usually disposed of in an underground system that passes the effluent through carefully selected undisturbed soil profiles. These soils must further filter and remove nutrients as the effluent makes its way to the groundwater or other bodies of water.

One variation of this type of treatment uses part of its treated effluent for toilet flushing. The stored recycled effluent should receive some aeration to ensure odour-free recycle water in the toilets.

This type of waste treatment system has several advantages. It is relatively inexpensive to install, and is not complex to operate and maintain. It provides a consistently good quality effluent. Most components can be installed aboveground, or they can be buried or partly buried. Good quality flushing water can be provided to reduce water supply needs and final disposal volumes.

The disadvantages are that septic tank odours must be dealt with properly. Several pumps and blowers can be involved, which creates maintenance and power costs, and can cause noise in some developments. Septic tank solids require proper disposal. An aerobic digestion tank would reduce volumes for disposal. A sand drying bed can be used for digested sludge drying, and the dried sludge should be buried in a landfill. A fairly large area is required for installation. There is no known manufacturer of a complete package system.

Prior to committing time and resources, it is prudent to contact actual users of similar systems-talk with a designer, system operator, an owner, and possibly a regulatory agency inspector who has observed performance.

Aerobic Waste Treatment

Aerobic waste treatment is also accomplished with microorganisms. However in this system, air is bubbled into the treatment process to ensure plenty of free oxygen. Aerobic organisms work about 20 times faster than anaerobic organisms. No offensive gas odours are released in this treatment process. Since they process is so much faster, much less holding time is required and less treatment area is needed.

This type of treatment plant is reasonable to consider for all sizes of park and tourism development, and because offensive odours are not normally produced, this type of system can be located close to occupied areas. The normal effluent quality of a properly designed and operated system meets most all secondary treatment standards With a reasonably small filter, high quality effluent can be produced for irrigation and recycled toilet flushing water. With carefully planned effluent recycling and irrigation reuse, little or no effluent disposal would be required. If disposal is required, several options are possible.

There are several manufacturers of small aerobic treatment plants. The best plants for harsh environments and isolated areas are fabricated from durable, low maintenance materials, simple to operate and maintain, designed to use the fewest moving parts, and consistent in effluent quality. They are also quick to install and put in service, proven in similar installations, and completely assembled and warranted by one manufacturer. A good manufacturer also provides, at a reasonable cost, technical and monitoring services on the new plant for at least one year after initial startup.

Advantages of such a system are that all materials in contact with liquids are noncorrosive (plastic, fibreglass, rubber, stainless steel) and can be buried or installed aboveground. The blower is the only moving part and requires only simple, routine maintenance. Effluent quality is consistently good. Excellent recycle and reuse options are available. The process easily handles surge loadings, as well as underloaded periods. Land area required is about 0.1 acre per 10,000 gallons of rated capacity. Site installation and startup can usually be accomplished in five working days or less. Few spare parts are needed. Purchase costs are competitive with other similar-sized treatment units. Unskilled operators have successfully followed infrequent, telephoned system adjustment instructions to maintain top performance.

The disadvantages are that most components are fabricated, assembled, and shipped from Colourado-few local materials would be used. The blower, although small, needs to be run continuously. The noise would have to be muffled, and a constant power cost would be involved.

The National Park Service has five of these plants in service in four widely varying environmental areas. One plant has been in service for several years.

Alternative Disinfection Methods

Traditionally, water from conventional treatment systems is disinfected with chlorine or chlorine compounds before being released back into the environment or reused. A side effect of this is that the chlorine or chlorine compounds are very reactive and sometimes produce highly persistent, toxic chemicals. Many environmentalists believe that there is no justification for use of chlorine and its compounds for disinfection.

Most public health codes call for disinfection with chlorine, and they would have to be changed to allow for either no disinfection or the use of other disinfectants. Less desirable possibilities include using less chlorine or removing the chlorine after the proper contact time. Dechlorination requires additional chemical feed and training.

The purpose of disinfection is to ensure that no virulent organisms are present after the water has been processed in one of the systems described earlier. The most common alternative disinfectants are ozone and ultraviolet light.

An entirely new treatment technology, introduced in September 1993, appears to provide excellent disinfection without the formation of environmentally harmful by-products. The National Park Service is arranging pilot studies at existing water and sewage treatment facilities to evaluate this emerging innovative technology.

❐

8

Nutrient Cycles

The main flow of energy that helps shape conditions on Earth's surface comes from space; and when the energy's work here is done, to space it returns. With respect to energy, then, Earth is an open system. With respect to its chemical endowment, however, Earth is a closed system. That is, the amounts of carbon, hydrogen, oxygen, iron, gold, and other elements in the planetatmosphere system do not change with time; the chemical arrangement and physical distribution of these elements can and do vary, but essentially nothing enters and nothing leaves the system.

The elements in this closed system that are essential to life are called nutrients. They can be divided into three categories: the 4 main chemical building blocks of living matter (carbon, oxygen, hydrogen, nitrogen); 7 macronutrients, of which smaller but still significant quantities are required for life; and 13 micronutrients, or trace elements, of which tiny quantities perform essential functions. The suitability of any terrestrial or aquatic environment for the support of life depends on the availability of nutrients in appropriate forms and quantities. The processes that govern this availability (or lack of it) are known collectively as nutrient cycles, because of the way the individual basic stocks of physical material move cyclically through the living and non-living parts of the physical world.

These cycles function in support of life not merely by making nutrients continuously available—in other words, by maintaining the fertility of the environment -but also by limiting the accumulation of material in quantities, forms, and places in which it would damage organisms. It is important to recognize in this connection that the same elements and compounds that serve as nutrients for some organisms in some concentrations are often toxic for other organisms, or even for the same organisms at higher concentrations. For example, molecular oxygen (O2) is toxic to anaerobic organisms and, at high enough concentrations, even to mammals. Ammonia (NH3) is an important source of nutrient nitrogen for many plants but is toxic to people. Hydrogen sulfide (H2S) is a nutrient for certain types of bacteria but is extremely toxic to mammals.

DYNAMICS OF NUTRIENT CYCLING

The steps that make up all nutrient cycles involve two basic processes: physical transport and chemical transformation. They are:

- The hydrologic cycle, including evaporation and precipitation as well as the flow of streams, groundwater, and ice;
- Winds;
- Ocean currents;
- Geologic movement, especially the upward and downward motions at the boundaries of tectonic plates and geologic uplifting on continents.

A fifth agent of physical transport that is important in a few circumstances is the movement of organisms. For example, fish-eating birds that deposit their excrement on land provide a significant pathway by which nitrogen and phosphorus are transferred from the sea to the land. Fish such as salmon, that feed mostly in the ocean but migrate up freshwater rivers to spawn and die, perform a similar function, as do the ocean-caught fish consumed by continentdwelling human beings.

How the various mechanisms of transport link the principal "compartments" of the physical world. Obviously, the ease (and, hence, the rate) with which a given substance can move through these pathways depends on its physical state (solid, liquid, gas), on other physical properties (solubility, volatility), and on the possibility of chemical transformations that can change those properties. Such chemical transformations take place in living organisms and in the non-living environment. They are considered in more detail below. Each of the compartments indicated often is divided into subcompartments for the purpose of tracing nutrient flows. Thus, for example, it is customary to consider three subcompartments of material in both the soil and the oceans: living organic matter, dead organic matter, and the inorganic medium itself.

These subcompartments can be further subdivided—for example, organic matter, into plants, animals, and microorganisms; and seawater, into the wellmixed surface layer and the stratified deeper layers. Finally, within each subdivision of each subcompartment, the magnitudes of the stocks, or pools, of the important elements and compounds are of interest, as are the magnitudes of the flows into and out of the pools (where flows can mean physical movement or chemical transformation or a combination of the two). For any pool that is in steady state—that is, for which inflows balance outflows so the size of the pool does not change with time—an average residence time for a compound can be calculated as the ratio of the size of the pool to the sum of the inflows or outflows. Examination of the residence times for important pools provides quick insight into the dynamics of nutrient cycling, and, indirectly, into the probable vulnerability of specific components of nutrient cycles to disruption by human intervention. For example, a long residence time means the annual flows are small compared to the pool, which means in turn that unless human disruptions cause a very large alteration in inflow or outflow they will affect the size of the pool only very slowly. A short residence time, by contrast, means that significant changes in inflow or outflow can affect the size of the pool very quickly.

CHEMISTRY OF NUTRIENT CYCLES

Two kinds of chemistry play crucial roles in the principal global nutrient cycles: the chemistry of the sedimentary cycle, associated with the very slow transformation of very large pools of material, and the chemistry of life, associated with the more rapid transformation of much smaller pools. Both kinds of chemistry depend very heavily on the very special roles of water as a source of hydrogen ($H+$) and hydroxyl ($OH-$) ions for many chemical reactions, as a medium for holding other chemical compounds in solution where they can react, and as a vehicle for physical transport of compounds from place to place.

Sedimentary Cycle

Nutrients are mobilized in the sedimentary cycle when water comes in contact with rock. Rainwater is slightly acidic, having absorbed some atmospheric carbon dioxide to form weak carbonic acid:

$$H_2O + CO_2 \longrightarrow H_2CO_3$$

$$H_2CO_3 \longrightarrow H^+ + HCO^-_3$$

The pH of rainfall in areas free of industrial pollution is around 5.7. (Recall from basic chemistry that pH is the negative of the log to base 10 of the concentration of hydrogen ions in moles per litre. The pH of distilled water is 7.0, corresponding to 10–7 moles of hydrogen ion per litre.) This slight acidity is enough to facilitate greatly the dissolution of many minerals present in exposed rock—a process called weathering.

Among the most important weathering reactions is that of the common mineral feldspar (a constituent of granite):

feldspar carbonic acid water

$$2KAlSiO_{8+} + 2\,(H^+ + HCO^-_3) + (H_2\,O) \longrightarrow$$

kalinite dissolved silica dissolved potassium andbicarbonate ion

$$Al_2Si_2O_5(OH)_4 + 4\,SiO_2 + 2K^+ + 2K^+ + 2K^+ + 2HCO_3$$

Kaolinite is a clay; it is left behind in this process as a contribution to the formation of soil, whereas the dissolved silica, potassium ion, and bicarbonate ion are carried away in the runoff. The mobilization and removal of positive ions such as (in this case) potassium is called leaching. Note that the reaction consumes some water and that it reduces the acidity of the solution (that is, it removes hydrogen ions) as it proceeds. These properties are characteristic of weathering reactions. Essentially the same reaction also occurs in the weathering of the aluminosilicates of sodium and calcium ($NaAlSi_3O_8$, $CaAl_2Si_2O_8$). Other silicate minerals also weather in the same general manner to produce kaolinite or other clays.

The weathering of certain other minerals leads to their complete dissolution; no solid trace is left behind to be incorporated into soil. The most important of these minerals are the limestones, or carbonate rocks, which undergo the following reactions:

$$CaCO_3H^+ + HCO_3^- \rightarrow Ca^{++} + 2HCO3^-$$

$$CaMg(CO_3)_2 + 2(H^+ + HCO3^-) \rightarrow Mg^{++} + Ca^{++} + 4HCO_3^-$$

Still other weathering reactions consume atmospheric oxygen in the oxidation of iron minerals:

$$4FeSiO_3 + O_2 + 2H_2O \rightarrow 4FeO(OH) + 4SiO_2$$

$$4Fe_3O_4 + O_2 \rightarrow 6Fe2O3$$

$$4FeS_2 + 15O_2 + 8H_2O_2 \rightarrow Fe_2O_3 + 8H_2SO_4$$

Most of the products of the weathering reactions mentioned here eventually reach the sea, either as dissolved ions or as suspended solids. Between the time they are freed from rock and the time they reach the sea, of course, some of these elements have made one or more passages through the living part of the biosphere as nutrients. In the sea, too, the nutrient substances are captured from the water by living organisms, eventually to be returned to the water or to the sediments on the ocean floor.

The reverse of many of the weathering reactions also can occur in seawater in the absence of intervention by living organisms, forming solid precipitates that join the organic debris on the ocean bottom.

Apparently these outputs of the major nutrient elements from the oceans roughly balance the inputs, because considerable physical evidence indicates that the salinity of seawater has been approximately constant for at least the past 200 million years.

The sedimentary cycle is finally closed by the geologic processes that is, by the processes of transformation and transportation that turn seafloor sediments into continental rocks, and by volcanic action.

Living Matter

Ecologist Edward S. Deevey, Jr., has pointed out that living matter can be regarded as a chemical compound with the empirical formula

$$H_{2960}\, O_{1480}\, C_{1480}\, N_{16P1.8}\, S.$$

This enormous (and completely intentional) oversimpli-fication serves both to call attention to the chemical side of life the side concerned -and, more specifically,the relative proportions in which the six most abundant elemental constituents of living material exist.

Actually, there is substantial variation in the ratios of the chemical elements of life in different kinds of organisms, the above formula being a weighted average in which woody plants predominate. The numbers are, in any case, not known with great accuracy. The basic reactions that link the water cycle, the carbon dioxide cycle, and the oxygen cycle through living matter are photosynthesis:

$$6CO_2 + 12H_2O + \text{solar energy} \rightarrow 6\, O_2 + C_6H_{12}O_6\, 6\ + H2O$$

and its opposite, respiration, or oxidative metabolism:

$$6O_2 + C_6H_{12}O_6 \rightarrow 6CO_2 + 6H_2O\ \text{heat}$$

The compound ($C_6H_{12}O_6$, or $(CH_2O)_6$, is glucose, the simplest carbohydrate produced by photosynthesis. The photosynthesis reaction is written with what appear to be six extra H_2O molecules on each side because twelve molecules of H_2O are actually broken up for each molecule of $C_6H_{12}O_6$ synthesized; radioactive-tracer experiments show that all twelve oxygen atoms from the 12 H_2O end up in the $6O_2$, and the oxygen in the six new water molecules comes from the CO_2. The amount of energy transformed from light into stored chemical energy by the photosynthesis reaction is 112 kilocalories (469 kilojoules) per mole of carbon fixed. The net result of the respiration reaction is to transform the same 112 kilocalories per mole of carbon from chemical energy into heat. Photosynthesis is carried out not only by higher green plants (such as oak trees, grasses, and ferns) but also by various kinds of algae (green, brown, red, blue-green) and by certain bacteria. In the bacteria, substances other than water serve as the sources of hydrogen for the reduction of carbon. For example, the green sulfur bacteria use hydrogen sulfide instead of water, according to the equation

$$12H_2S\,6 + CO_2 \rightarrow 6CO_2 \text{ light} \rightarrow C_6H_{12}O_6 \rightarrow H_2O + 12S$$

producing elemental sulfur rather than oxygen. (In fact, these bacteria are obligate anaerobes—they would be poisoned by oxygen.) The reverse reaction (respiration) consumes sulfur and carbohydrate and liberates heat, carbon dioxide, and hydrogen sulfide. Still other photosynthesis reactions use hydrogen, obtained from H_2O or H_2S or other hydrogen donors, to reduce substances other than carbon dioxide. For example, most higher plants can reduce nitrate, instead of CO_2, in photosynthesis:

$$9H_2O + 2NO_3 + \text{light} \rightarrow 2NH_3 + 6H_2O\,4\,O_2.$$

and nitrogen-fixing photosynthetic organisms can reduce atmospheric nitrogen:

$$6O2 + C_6H_{12}O_6 \rightarrow 6CO_2 + 6\,H_2O \text{ heat}$$

These variations notwithstanding, the great bulk of photosynthesis on Earth consists of the reduction of carbon dioxide, with water as the hydrogen source.

The photosynthesis and respiration reactions, as we have written them here, actually represent only the end points of vastly more complicated sequences of chemical events. For example, although burning glucose in air is described essentially completely by the respiration reaction given

$$6O_2 + C_6H_{12}O_6 \rightarrow 6\,CO_2 + 6H_2O + \text{heat},$$

to achieve the same result in a living cell (at a temperature far below that of combustion) requires many intermediate steps. The intermediate steps in both respiration and photosynthesis rely heavily on three kinds of organic compounds: the energy carrier ATP (adenosine triphosphate); the electron carrier NADP (nicotinamide adenine dinucleotide phosphate); and various catalysts called enzymes. (Catalysts speed up chemical reactions without themselves being consumed.)

A closer look at these compounds provides some clues to the importance of the elements nitrogen, phosphorus, and sulfur in the chemistry of life. All biochemical reactions of importance (not just photosynthesis and respiration) require enzyme catalysts if they are to proceed at significant rates. All enzymes are proteins. Proteins other than enzymes also serve as hormones; as vehicles to transport oxygen; as important structural components of skin, tendons, muscle, cartilage, and bone; and in numerous other roles. All proteins are made up of building blocks called amino acids, and all amino acids contain the amino group—NH_2. A single protein contains from fifty to some tens of thousands of amino acids, and hence at least that many nitrogen atoms. (Some amino acids contain more than one nitrogen.) The shape and stiffness of many proteins—properties essential to their functions—are governed and maintained by bonds in which sulfur plays a crucial role, so it, too, is essential to the chemistry of life.

ATP and NADP are examples of nucleotides, compounds consisting of a five-carbon sugar to which one or more phosphate groups ($–PO_4$) and a nitrogen base are attached. A nitrogen base is a ring compound containing both nitrogen and carbon in the ring structure. Nucleotides not only perform critical energy transfer and electron-transfer functions in the chemistry of living cells, but they are also the building blocks of nucleic acids. Each nucleic acid molecule contains hundreds to thousands of nucleotides, the arrangement of which is a chemical means of storing information. Nucleic acids control the synthesis of proteins and a variety of other processes in living cells. The nucleic acid DNA (deoxyribonucleic acid) is the genetic material—the bearer of the coded information that is passed from parent cells and organisms to their offspring to "tell" them what to become. The nucleic acid RNA (ribonucleic acid) plays essential roles in carrying out the instructions coded in DNA molecules.

In addition to its role as a constituent of DNA and RNA, phosphorus, in the form of phosphate, is one of the principal anions (ions with a negative charge) that maintains electric neutrality in the fluids of living organisms. Along with calcium, it is also an important component in shells and bony structures. Sulfur, besides playing its role in proteins, in the form of sulfate (SO_4) is another principal anion in body fluids and cells. The third principal anion is the chloride ion (Cl). The principal cations (positively charged ions) are sodium, potassium, calcium, and magnesium. Magnesium is an essential constituent of many enzymes and of the green pigment chlorophyll. Iron is a key component of many enzymes and of the oxygen-transporting protein hemoglobin.

Oversimplifying considerably, one can say that the quantity of living matter in a given environment is limited by the stock of that requisite of life that is in shortest supply. This rather obvious proposition carries the name Liebig's law of the minimum. Of course, the limiting factor varies from one type of organism to another and from one environment to another.

Generalizations are difficult, but it is fair to say that under many circumstances the quantity of plant material in a given environment is in fact controlled by a limiting nutrient (although the density of a given kind of plant is often controlled by a herbivore). In desert regions the limiting nutrient is generally water. Under many conditions where water is not limiting, as in freshwater lakes, the limiting nutrient turns out to be phosphorus; in the open ocean it may be iron. In some richly productive environments, nitrogen and phosphorus are jointly limiting—producing any more plant material would require supplying more of both nutrients. Obviously, the limiting nutrient in a given biological system represents a leverage point where the effects of natural or human-induced perturbations manifest themselves rapidly.

CYCLES OF THE PRINCIPAL NUTRIENTS

The cycles of many nutrients are tightly linked chemically and biologically. This is so much the case for the cycles of carbon, oxygen, and hydrogen that we consider them together. The cycles of nitrogen, phosphorus, and sulfur, on the other hand, can usefully be examined separately.

Carbon-Oxygen-Hydrogen

It is appropriate to begin our closer examination of the carbon-oxygenhydrogen cycles with their intersection in the green plant, the basis of almost all life on Earth.

Carbon and Energy in Living Plants

Most living plants not only fix (store) energy by means of photosynthesis but they also use some of that stored energy to drive their own life-sustaining internal processes by means of respiration.

The total rate at which a plant or a plant community stores solar energy by means of photosynthesis is called gross photosynthesis or gross primary production (GPP). The figure obtained by subtracting from gross photosynthesis the amount of energy used by the plants in respiration (R) is called net photosynthesis or net primary production (NPP). In abbreviated form, then, the relation is GPP-R = NPP. Several interchangeable units are used in the biological literature for the measurement of these quantities: calories or kilocalories per unit of time, watts or kilowatts (different units for measuring the same thing—energy flow), grams of carbon fixed into organic compounds per unit of time, moles of carbon fixed per unit of time, or grams of dry organic matter produced per unit of time.

A wide range of estimates of global net primary production and the associated standing crop of living plants (phytomass) have been published. One recent review indicates that the estimates of net primary production are converging on a value around 160 billion (1.60 x 10) metric tons per year for the globe.

Global phytomass was estimated in the same review as about ten times larger 1840 billion metric tons, dry weight and gross photosynthesis was estimated to be about twice as large as net production (that is, half of gross photosynthesis is consumed in plant respiration and half remains as net production). The estimates for phytomass and net primary production, separated into terrestrial and oceanic components. The roughness of these estimates deserves emphasis. The range of recent estimates of land phytomass 30 per cent larger; the range of published values of ocean phytomass is from twenty times smaller than the value to three times larger. The factor-of-60 range of informed opinion with respect to oceanic phytomass is especially remarkable.

The range in recent published estimates of oceanic net primary production is only ± 10 per cent from the value, and the range in terrestrial net production is from 15 per cent less to 60 per cent more than the value in the table. The fraction of gross primary production consumed in plant respiration is known to vary between 20 per cent and 75 per cent, depending on type of plant and region, but the global average is unlikely to differ much from 50 per cent.

Energy and Material Flow in the Carbon Cycle

For the purpose of reconciling estimates of annual production presented in grams of carbon, grams of dry organic matter, and kilocalories, one widely used convention assumes that 10 kilocalories of energy are stored per gram of fixed carbon and that dry organic matter contains 45 per cent fixed carbon by weight. Thus, measurements in grams or metric tons (1 MT = 106 g) of carbon can be converted to grams or metric tons of dry organic matter by multiplying by 2.2, and 1 gram of dry organic matter can be taken to represent 4.5 kilocalories of stored energy. Useful conversions following directly from this convention are that I gram of dry organic matter per day corresponds to an average energy flow of about 0.2 watts; 1 gram of carbon per day, to about 0.5 watts; and one metric ton of dry organic matter per year, to about 600 watts. Also, 1 metric ton of dry organic matter contains 37,500 moles of carbon.

Although we shall use the convention just described, the reader may encounter others elsewhere. If the dry organic matter were pure carbohydrate (CH_2O)n which stores 112 kilocalories per mole of carbon, then the appropriate conversions would be: 3.7 kilocalories per gram of dry organic matter, 9.3 kilocalories per gram of carbon, 0.4 gram carbon per gram of dry organic matter [Actually, CH_2O is a good approximation of the average composition of all living matter because so much of the world's biomass is cellulose tied up in woody plants, and cellulose has nearly this formula. However, most of the plant material produced each year is not cellulose—remember, a big pool does not necessarily mean a big flow. The annual production of organic matter contains a higher proportion of fixed carbon and higher energy content per gram than does the cellulose-dominated, slow-toturn-over standing crop.] The rounded-off value of 4 kilocalories per gram of dry organic matter is in fairly widespread use in conjunction with the figure of 10 kilocalories per gram of fixed carbon, itself a rough average among several kinds of plant matter.

Taking the range of estimates into account, one can state several conclusions with reasonable assurance:

- The annual NPP of the oceans is only one-third to one-half that of the continents, even though the oceans cover more than 70 per cent of Earth's surface;
- The pool, or standing crop, of living plant material on land is at least several hundred times that in the oceans;
- The average turnover time (life expectancy) of oceanic plants must be in the range of days to weeks, in order for so small a phytomass to be associated with so large an annual production;
- The average turnover time of terrestrial living plants, by contrast, is in the range of a decade or two;
- Oceanic and terrestrial phytomasses are both so poorly known that man-induced or natural changes large enough to be very important in the long term could easily escape notice for some time. In the latter regard, although there are good reasons to suppose that the activities of civilization have diminished the terrestrial phytomass significantly, some experts consider this assertion to be debatable.

Notwithstanding the uncertainties in the data, it is instructive to examine the large regional variations in plant productivity that are concealed in the global totals. Of particular interest are the facts that tropical forests account for more than 40 per cent of terrestrial NPP (although they cover only about 16 per cent of the land area) and that up- welling regions, continental shelves, reefs, and estuaries yield almost 25 per cent of oceanic NPP while accounting for less than 9 per cent of the ocean's area.

The rate at which energy is being stored in net primary production is about 100 million megawatts, roughly two-thirds on land and one-third in the ocean. The rate of gross primary production can be assumed to be about 200 million megawatts. The net primary production is roughly twelve times the rate at which civilization used commercial energy in 1975, where commercial energy refers to energy that is sold (for example, fossil fuels, hydropower, and nuclear power, but excluding food) as opposed to energy that is gathered by the user (for example, most wood, dung).

Very little of the solar energy that reaches Earth's surface is captured by photosynthesis. The net primary production is about 0.1 per cent of the incident solar energy at the surface-0.25 per cent of what strikes the land and 0.05 per cent of what strikes the ocean. Some of the reasons these percentages are so small are obvious. Plants do not by any means cover all of Earth's surface; in many areas of the world the growing season lasts only part of the year; and in some circumstances, as already noted, plant growth is limited by lack of one or more critical nutrients. Under favourable conditions, on the other hand—as in a well watered and fertilized cornfield or a fertile swamp—NPP may reach 2 per cent or so of incident solar energy averaged over the growing season and as much as 6 per cent or 7 per cent of incident solar energy on the most favourable days.

One additional reason for this is that only about one-fourth the solar energy reaching the ground is in the part of the wavelength range that stimulates photosynthesis (blue and red light). The other big reason is the large amount of water that must be evaporated through the leaves of plants to maintain the nutrient-bearing flow of water through them. Evaporating this water typically takes 40 per cent or more of the energy that falls on a plant, and a much higher percentage of what is actually absorbed by the plant. (For production of every ton of wet weight of many crops—of which about onefourth is dry organic matter around 100 tons of water are evaporated. This means 400 grams of water are evaporated, requiring about 0.6 kilocalories per gram evaporated, for every gram or 4.5 kilocalories of dry organic matter produced. The ratio in such a case is about 50 kilocalories for evaporation per kilocalorie of net primary production.)

OXYGEN BALANCE: METABOLISM BY ANIMALS AND MICROBES

The overall effects of the net production of plants on the cycles of carbon, oxygen, and hydrogen are as follows:

- More oxygen, O_2, is added to the environment than is removed from it;
- Less carbon dioxide, CO_2, and water, H_2O, are added to the environment than are removed from it;
- The CO_2 and H_2O subtracted from the environmental pools show up as additions to the pool of stored organic matter, CH_2O, and to the atmospheric oxygen pool. Quantitatively, for every metric ton of dry organic matter produced, there is a net loss from the environment of 0.6 metric tons of water and about 1.5 metric tons of CO_2, and a net gain of about 1.1 metric tons of oxygen.

Why, then, is oxygen not building up continuously in the atmosphere, while water and carbon dioxide are depleted?

Because plant matter is not accumulating, but rather is approximately in steady state. It is being broken down into carbon dioxide and water again by the metabolism of animals and microbes, using up molecular oxygen just as rapidly as new plant material and oxygen are being produced.

Consider the possible fates of plant carbohydrate produced as part of net primary production. Plants either are consumed by herbivores, or they die and add their carbohydrate to the pool of dead organic matter. Of the part consumed by herbivores, some is metabolized and thereby returned to the environmental pool of CO_2 and H_2O, some is added as excrement to the pool of dead organic matter, and some is incorporated into herbivore tissue, either to be eaten by carnivores or to be added at the herbivore's death (by causes other than being devoured) directly to the pool of dead organic matter. The same fates await the carbohydrate that reaches the carnivores. The cycle is closed by the action of decomposers—both animals and microbes (bacteria)— which by metabolizing the fixed carbon in dead organic matter return CO and H O to the environmental pools.

There are some important side pathways in the carbon cycle in which the initial products of decomposition include methane and carbon monoxide (CH_4 and CO) in addition to carbon dioxide. The main reactions are:

$$2CH_2 + O \rightarrow CH_4 \; CO_2$$

$$2CH_2O + O_2 \rightarrow 2CO + 2\,H_2O$$

The CH_4 and CO may then be oxidized in the soil or in the atmosphere according to the following reactions:

$$CH_4\; 3\,O_2 \rightarrow 2CO + 4\,H_2O$$

and

$$2CO + O_2 \rightarrow 2CO_2$$

The details concealed by these simple formulas can be quite complex. In steady state, the metabolism of carbohydrate by herbivores, carnivores, and decomposers returns to the environment exactly as much CO_2 and H_2O as the production of the carbohydrate by photosynthesis originally removed. And it consumes exactly as much oxygen as the photosynthesis produced. There is therefore no long-run buildup (or depletion) of oxygen in the environment as a result of this cycle, unless net photosynthesis and metabolism get out of balance.

THE SEQUESTERING OF FIXED CARBON

How did oxygen accumulate in the atmosphere in the first place? The biogeochemical evidence suggests that molecular oxygen in the atmosphere is of biological origin, there having been essentially none before the evolution of photosynthetic organisms, perhaps 2 billion years ago. For oxygen produced by photosynthesis to have accumulated over the long term, part of the carbon fixed by this process must have been withdrawn from the cycle, escaping the oxidation step back to carbon dioxide. They originated as plant material that escaped breakdown by decomposers, in many cases by being buried in swamps and bogs under conditions in which both oxygen and anaerobic decomposer organisms were absent. Hundreds of millions of years of burial under considerable pressure and heat converted these plant materials into the hydrocarbon fossil fuels—coal, petroleum, and natural gas. The oxygen released when the carbon in these fuels was first fixed by photosynthesis remained in the atmosphere.

The different ranks of coal—lignite, subbituminous, bituminous, anthracite—represent different ages (listed here from youngest to oldest) in terms of the origin of the plant material they derived from, and somewhat different chemical compositions. Peat is the partially decomposed organic matter that is the precursor of coal. Petroleum, natural gas, oil shale, and tar sands are the results of different histories of biogeochemical transformation, but all originated, like coal, with dead organic matter. Not all of the fixed

carbon that has been sequestered out of reach of natural oxidative processes is in deposits as concentrated as the fossil fuels. A much larger amount, in fact, is dispersed in sediments whose hydrocarbon content is on the order of 1 part per 1000 (1 g of hydrocarbon material per kg of sediment) or less.

Since the oxygen reservoir in the atmosphere has been built up by the sequestration of fixed carbon in various forms, could it be jeopardized by any forseeable activities of civilization that would cause the oxidation of these fixed-carbon pools? Deforestation, for example, could lead to the oxidation of part of the standing crop of fixed carbon in living organic matter, and combustion of fossil fuels oxidizes those pools. Yet, examination of the best estimates of the sizes of the fixed-carbon pools shows little cause for concern in the short term. Removal and oxidation of the entire terrestrial biomass would deplete atmospheric oxygen by at most 0.2 per cent. Combustion of all the conventional fossil fuels (coal, petroleum, natural gas—of which about 90 per cent is coal) thought to be recoverable ever by human effort would deplete atmospheric oxygen by 1.8 per cent. This would reduce the oxygen content of the atmosphere at sea level to what it now is at an elevation of 150 meters.

The 1975 global rate of fossil-fuel combustion would amount to a depletion of atmospheric oxygen by about 0.001 per cent per year—far below the threshold of detectability with existing measurement techniques. On the other hand, the amount of reduced carbon thought to be present in oil shales is sufficient, if it were ever totally recovered and oxidized, to consume more than half the atmospheric oxygen pool.

Some observers believe that most of the oil shales are too dispersed an energy resource ever to be economically attractive. But, in the event this view is not correct and civilization commences to use oil shale at a very high rate— say, ten times the rate at which all forms of energy were consumed world wide in 1975—it would still take many centuries for the resulting depletion of atmospheric oxygen to become serious.

The oxygen situation is made more complicated than indicated in the foregoing paragraphs by the existence of important pools of oxygen elsewhere than in the atmosphere. It is apparent from the information that other oxygen pools must exist, because the number of moles of oxygen in the atmosphere is much smaller than the number of moles of reduced carbon in the sediments and elsewhere.

The oxygen produced when all that carbon was fixed and sequestered was tied up with the oxidation of ferrous oxide (FeO) to ferric oxide (Fe_2O_3) and sulfur to sulfate ($-SO_4$) in sedimentary rocks. It may also have been tied up in the oxidation of carbon of inorganic origin (such as carbon monoxide from volcanoes) to carbon dioxide and carbonate ($-CO_3$). Although all these reactions are slow, they can account for the necessary enormous amounts of oxygen if they are assumed to have operated over geologic time spans. Similarly, alterations in conditions governing the reaction rates could lead to significant changes in atmospheric composition over very long periods. On shorter time scales, however, organic carbon pools and associated reactions are almost certainly more important, and they are unlikely to produce significant changes in atmospheric oxygen in less than centuries.

CARBON DIOXIDE

Although atmospheric oxygen depletion evidently is not a short-term threat, buildup of atmospheric carbon dioxide is. Every mole of oxygen removed from the atmosphere means a mole of carbon dioxide added. Since the number of moles of CO in the atmosphere is 700 times smaller than the number of moles of O, a change of a specified number of moles has 700 times the effect on the CO concentration as on the O, concentration. The combustion of fossil fuel that depletes oxygen by 0.001 per cent per year augments the atmospheric CO by 0.7 per cent per year—not an insignificant increment.

Not all of the added CO remains in the atmosphere, however. Some is absorbed by the oceans, and it is conceivable that some is being taken up in an increase in biomass. (Recall from the earlier discussion that the size of the biomass is not known with sufficient accuracy to detect even rather substantial changes.) An increase in the concentration of CO_2 in the atmosphere should increase the rate of net photosynthesis (where other nutrients are not limiting), and this would produce a growing biomass if the rate of oxidative metabolism by consumers and decomposers did not increase correspondingly.

Accurate measurements of atmospheric CO_2 have been made since about 1957. They have been coupled with indirect evidence to produce a model of the rise of atmospheric CO2 from about 280 molecular parts per million in preindustrial times to around 325 parts per million in 1975. An additional kind of evidence that can shed some light on the sources and fate of the CO_2 added to the atmosphere is the abundance of radioactive carbon-14 relative to the amount of non-radioactive carbon-12 in various pools. Carbon-14 is produced naturally by bombardment of nitrogen by cosmic rays in the stratosphere and also by civilization when nuclear bombs are detonated in the atmosphere; it then mixes throughout the atmosphere and is incorporated, along with other carbon, into plant matter.

Since the half-life of carbon-14 is only 5770 years, the carbon-14 buried for millions of years in fossil fuels has decayed away entirely. Thus, the CO_2 added to the atmosphere by the combustion of fossil fuels contains no carbon- 14, whereas CO added from pools with faster turnover does contain carbon- 14. The reduction in the 14C/C ratio in the atmosphere because of the combustion of fossil fuels is called the Suess effect. The use of data on the Suess effect, combined with other information needed to help interpret it (such as the way the rate of natural and artificial 14C production has varied over time), has allowed development of the estimate that about half of the CO ever added to the atmosphere by combustion of fossil fuels has stayed there. The same work indicates that the land biomass should have increased between 1per cent and 3 per cent since preindustrial times, having absorbed some of the industrial CO_2, while the oceans absorbed the rest. This change in land biomass is much too small to measure, and the exact amount is a matter of controversy.

GLOBAL CARBON CYCLE

The pools and flows of the global carbon cycle. We caution the reader once more that substantial uncertainties exist about many of the numbers, as noted in detail earlier, and that further research might therefore alter the picture considerably. The pools of living and dead organic matter on land and in the oceans are assumed to be in equilibrium. Half the CO_2 added to the atmosphere by combustion of fossil fuels is assumed to dissolve in the surface layer of the oceans.

The carbon added to the atmosphere by the respiration of animals and microbes, only a few per cent is in the form of methane (CH_4) and carbon monoxide (CO), while the rest is carbon dioxide. Still, the natural sources of carbon monoxide—incomplete decomposition of organic matter and partial oxidation of methane in the atmosphere—exceed man-made carbon monoxide from fossil-fuel combustion severalfold. Methane is apparently not toxic at any concentrations that occur in natural environments, and carbon monoxide is toxic only at concentrations on the order of 100 times the global atmospheric average or more.

NITROGEN

The nitrogen cycle is chemically the most intricate of the major nutrient cycles, and it is probably the least well understood scientifically. The crucial role of nitrogen in all proteins makes this element essential to life, but most organisms can assimilate and use nitrogen only in specific chemical forms. Only a relatively few species of bacteria and algae can convert gaseous molecular nitrogen (N_2) into the more complex compounds that can be used by plants and animals.

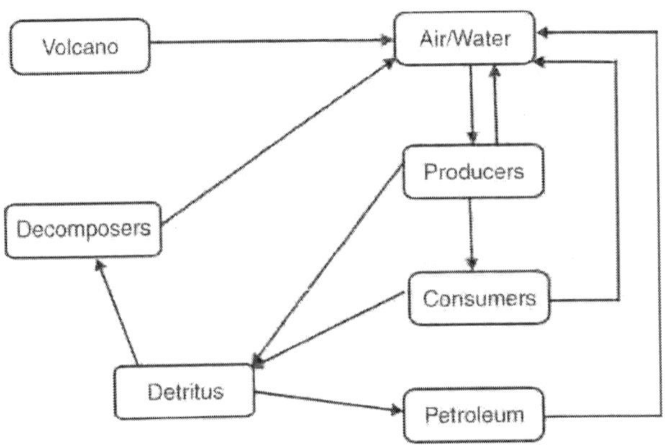

Fig. 8.1: Pools and Flows in the Global Carbon Cycle

The principal usable forms are ammonia (NH_3) and nitrates ($-NO_3$), of which the latter is needed in greater quantity by most plants. The conversion of N2 to ammonia and nitrate is called nitrogen fixation, and the species that can accomplish this conversion are called nitrogen-fixing organisms. Thus, although the atmosphere represents an enormous reservoir of molecular nitrogen, the continuation of life on Earth depends absolutely on the activities of these tiny, inconspicuous organisms.

ORGANISMS AND NITROGEN

The best known of the nitrogen-fixing organisms are the bacteria associated with the special nodules on roots of legumes, which are plants of the pea family. In a symbiotic relationship, the bacteria obtain from the plants the energy they need to carry out the nitrogenfixing reaction;

$$2N_2 + 6H_2O \rightarrow 4NH_3 + 3O_2$$

and some of the ammonia, in turn, is made available to the plants for the synthesis of amino acids, for example,

$$2NH_3 + 2H_2O + 4CO_2$$

$$2CH_2\,NH_2\,COOH + 3O_2.$$

The capacity of such legumes as alfalfa, beans, peas, and clover to fix nitrogen in large quantities has led to their widespread use in crop-rotation schemes to replenish in the soil the nitrogen depleted by other crops.

Although nitrogen fixation by the legume-bacteria symbiosis is thought to be responsible for most natural nitrogen fixation, the fixation step is also carried out by some bacteria that have looser associations with plants, including the free-living Azotobacter (aerobic) and Clostridium (anaerobic), and by various species of blue-green algae. Blue-green algae are apparently responsible for maintaining the fertility of rice paddies in much of Asia, which are subjected to intensive cropping without the use of nitrogen fertilizers.

The amino acids used by all organisms as the building blocks of protein are synthesized not only from ammonia but also from nitrate produced from ammonia by processes collectively called nitrification. The nitrification reactions yield energy, which supports the life processes of the nitrifying bacteria (Nitrosomonas, Nitrobacter). The actual synthesis of amino acids is carried out by bacteria (which differ widely in how many amino acids they can synthesize), by plants (which can synthesize all the amino acids needed for making proteins), and by animals (which, again, differ in the variety of amino acids they can produce).

Vertebrates—including human beings—cannot synthesize all the amino acids they need and must obtain intact in the food they eat the ones they cannot synthesize. The amino acids an animal must get from its diet are called essential amino acids; those it can synthesize are called nonessential. Fixed nitrogen that has been incorporated into organisms returns to the soil in animal wastes and in dead organisms (microbes, plants, animals) or parts of organisms (for example, leaves). Animal wastes are rich in urea—$(NH_2)_2$ CO—which is the principal product of the metabolism of proteins in many organisms.

The proteins of dead organisms are broken down into amino acids and other residues by bacteria and fungi of decay. The latter compounds—urea, amino acids, and other breakdown products of protein—are then converted into ammonia by yet another group of bacteria. This step is called ammonification. (The chemical reaction for the ammonification of an amino acid is just the opposite of aminoacid synthesis.)

So far we have described how nitrogen is fixed by living organisms and how the fixed nitrogen is further processed by them. A logical question at this point is: How is nitrogen unfixed? For without a mechanism for returning nitrogen to its molecular form (N_2), even the vast reservoir of this compound that exists in the atmosphere would have been depleted long ago; much of the nitrogen would be tied up as nitrates in soil, in the oceans, and in sediments.

The mechanism that has prevented this outcome by closing the atmospheric loop of the nitrogen cycle is the action of a relatively few groups of denitrifying bacteria that make their living converting nitrate to nitrous oxide (N_2O) and N_2. (These transformations release energy.) The nitrous oxide is reduced to N_2 by further bacterial action, or in the atmosphere by photochemical reactions. Another set of denitrifying reactions, also carried out by bacteria, transforms nitrate to nitrite and nitrite to ammonia. The chemistry of the main steps in the nitrogen cycle that are mediated by bacteria is summarized. For each reaction, the "energy" column indicates whether energy must be supplied to accomplish it ("in") or whether the reaction makes energy available ("out"). The part of the terrestrial nitrogen cycle directly associated with organisms is represented schematically. The oceanic cycle is substantially similar, relying both on bacteria and algae.

❐

9

Environment Policy and Environment Management System

Harmony between man and environment is the essence of healthy life and growth. Therefore, maintenance of ecological balance and a pristine environment has been of utmost importance to the Union Ministry of Power (MoP). NTPC being the leading organization under the ministry in the areas of power generation, has been taking various measures discussed below for mitigation of environment pollution due to power generation.

Driven by its commitment for sustainable growth of power, NTPC has evolved a well defined environment management policy and sound environment practices for minimising environmental impact arising out of setting up of power plants and preserving the natural ecology.

NATIONAL ENVIRONMENT POLICY

At the national level, the Ministry of Environment and Forests had prepared a draft Environment Policy (NEP) and the Ministry of Power along with NTPC actively participated in the deliberations of the draft NEP. The NEP 2006 has since been approved by the Union Cabinet in May 2006.

NTPC ENVIRONMENT POLICY

As early as in November 1995, NTPC brought out a comprehensive document entitled "NTPC Environment Policy and Environment Management System". Amongst the guiding principles adopted in the document are company's proactive approach to environment, optimum utilisation of equipment, adoption of latest technologies and continual environment improvement. The policy also envisages efficient utilisation of resources, thereby minimising waste, maximising ash utilisation and providing green belt all around the plant for maintaining ecological balance.

ENVIRONMENT MANAGEMENT, OCCUPATIONAL HEALTH AND SAFETY SYSTEMS

NTPC has actively gone for adoption of best international practices on environment, occupational health and safety areas. The organization has pursued the Environmental Management System (EMS) ISO 14001 and the Occupational Health and Safety Assessment System OHSAS 18001 at its different establishments. As a result of pursuing these practices, all NTPC power stations have been certified for ISO 14001 and OHSAS 18001 by reputed national and international Certifying Agencies.

POLLUTION CONTROL SYSTEMS

While deciding the appropriate technology for its projects, NTPC integrates many environmental provisions into the plant design. In order to ensure that NTPC comply with all the stipulated environment norms, various state-of-the-art pollution control systems/ devices as discussed below have been installed to control air and water pollution.

Electrostatic Precipitators

The ash left behind after combustion of coal is arrested in high efficiency Electrostatic Precipitators (ESPs) and particulate emission is controlled well within the stipulated norms. The ash collected in the ESPs is disposed to Ash Ponds in slurry form.

Flue Gas Stacks

Tall Flue Gas Stacks have been provided for wide dispersion of the gaseous emissions (SOX, NOX etc) into the atmosphere.

Low-NOX Burners

In gas based NTPC power stations, NOx emissions are controlled by provision of Low-NOx Burners (dry or wet type) and in coal fired stations, by adopting best combustion practices.

Neutralisation Pits

Neutralisation pits have been provided in the Water Treatment Plant (WTP) for pH correction of the effluents before discharge into Effluent Treatment Plant (ETP) for further treatment and use.

Coal Settling Pits/Oil Settling Pits

In these Pits, coal dust and oil are removed from the effluents emanating from the Coal Handling Plant (CHP), coal yard and Fuel Oil Handling areas before discharge into ETP.

DE & DS Systems

Dust Extraction (DE) and Dust Suppression (DS) systems have been installed in all coal fired power stations in NTPC to contain and extract the fugitive dust released in the Coal Handling Plant (CHP).

Cooling Towers

Cooling Towers have been provided for cooling the hot Condenser cooling water in closed cycle Condenser Cooling Water (CCW) Systems. This helps in reduction in thermal pollution and conservation of fresh water.

ASH DYKES AND ASH DISPOSAL SYSTEMS

Ash ponds have been provided at all coal based stations except Dadri where Dry Ash Disposal System has been provided. Ash Ponds have been divided into lagoons and provided with garlanding arrangements for change over of the ash slurry feed points for even filling of the pond and for effective settlement of the ash particles. Ash in slurry form is discharged into the lagoons where ash particles get settled from the slurry and clear effluent water is discharged from the ash pond. The discharged effluents conform to standards specified by CPCB and the same is regularly monitored. This has been envisaged for the first time in Asia which has resulted in progressive development of green belt besides far less requirement of land and less water requirement as compared to the wet ash disposal system.

ASH WATER RECYCLING SYSTEM

Further, in a number of NTPC stations, as a proactive measure, Ash Water Recycling System (AWRS) has been provided. In the AWRS, the effluent from ash pond is circulated back to the station for further ash sluicing to the ash pond. This helps in savings of fresh water requirements for transportation of ash from the plant.

The ash water recycling system has already been installed and is in operation at Ramagundam, Simhadri, Rihand, Talcher Kaniha, Talcher Thermal, Kahalgaon, Korba and Vindhyachal. The scheme has helped stations to save huge quantity of fresh water required as make-up water for disposal of ash.

Dry Ash Extraction System (DAES)

Dry ash has much higher utilization potential in ash based products (such as bricks, aerated autoclaved concrete blocks, concrete, Portland pozzolana cement, etc.). DAES has been installed at Unchahar, Dadri, Simhadri, Ramagundam, Singrauli, Kahalgaon, Farakka, Talcher Thermal, Korba, Vindhyachal, Talcher Kaniha and BTPS.

Liquid Waste Treatment Plants and Management System

The objective of industrial liquid effluent treatment plant (ETP) is to discharge lesser and cleaner effluent from the power plants to meet environmental regulations. After primary treatment at the source of their generation, the effluents are sent to the ETP for further treatment. The composite liquid effluent treatment plant has been designed to treat all liquid effluents which originate within the power station e.g., Water Treatment Plant (WTP), Condensate Polishing Unit (CPU) effluent, Coal Handling Plant (CHP) effluent, floor washings, service water drains etc.

The scheme involves collection of various effluents and their appropriate treatment centrally and re-circulation of the treated effluent for various plant uses. NTPC has implemented such systems in a number of its power stations such as Ramagundam, Simhadri, Kayamkulam, Singrauli, Rihand, Vindhyachal, Korba, Jhanor Gandhar, Faridabad, Farakka, Kahalgaon and Talcher Kaniha. These plants have helped to control quality and quantity of the effluents discharged from the stations.

Sewage Treatment Plants and Facilities

Sewage Treatment Plants (STPs) sewage treatment facilities have been provided at all NTPC stations to take care of Sewage Effluent from Plant and township areas. In a number of NTPC projects modern type STPs with Clarifloculators, Mechanical Agitators, sludge drying beds, Gas Collection Chambers etc have been provided to improve the effluent quality.

The effluent quality is monitored regularly and treated effluent conforming to the prescribed limit is discharged from the station. At several stations, treated effluents of STPs are being used for horticulture purpose.

ENVIRONMENTAL INSTITUTIONAL SET-UP

Realizing the importance of protection of the environment with speedy development of the power sector, the company has constituted different groups at project, regional and Corporate Centre level to carry out specific environment related functions.

The Environment Management Group, Ash Utilisation Group and Centre for Power Efficiency and Environment Protection (CENPEEP) function from the Corporate Centre and initiate measures to mitigate the impact of power project implementation on the environment and preserve ecology in the vicinity of the projects. Environment Management and Ash Utilisation Groups established at each station, look after various environmental issues of the individual station.

Environment Management During Operation Phase

NTPC's environment friendly approach to power has already begun to show results in conservation of natural resources such as water and fuel (coal, oil and gas) as well as control of environmental pollution. As already mentioned earlier, NTPC has chalked out a set of well defined activities that are envisaged right from the project conceptualisation stage so that during the entire life cycle of the power plant, NTPC is fully compliant with various environment regulations and a pristine environment and ecological balance is maintained in and around its power station and townships.Following is brief description of some of the measures taken during the operation phase of the stations. Performance enhancement and up-gradation measures are undertaken by the organisation during the post operational stage of the stations. These activities have greatly helped to minimise the impact on environment and preserve the ecology in and around its power projects. These measures have been enumerated as follows.

Monitoring of Environmental Parameters

A broad based Environment Monitoring Programmes has been formulated and implemented in NTPC. All pollutants discharged from the power plant such as stack emission, ash pond effluent, main plant effluent, domestic effluent and Condenser Cooling Water (CCW) effluent are monitored at the stipulated frequency at the source itself and at the points of discharge. In addition to the above, ambient air, surface water and ground water quality in and around NTPC plants are regularly monitored to assess any adverse impacts as a result of operation of the power plant.

On-Line Data Base Management

In order to have better control on pollution and to achieve effective environment management in and around NTPC stations, it is imperative to have an on-line, reliable and efficient environment information system on the plant operational and environmental performance parameters at all three levels i. e generating Stations, Regional Headquarters and Corporate Centre. In consideration of the above, a computerized programmes, namely "Paryavaran Monitoring System"-PMS, which could provide reliable storage, prompt and accurate flow of information on environmental performance of Stations was developed and installed in NTPC. This software facilitates direct transfer of environment reports and other environment related information from stations to the Regional Headquarters and Corporate Centre. The PMS has already been implemented at Corporate Centre, the

Regional Headquarters and most of the Stations. This system has helped in achieving continuous improvement in NTPCs environment performance through improved monitoring and reporting system by using the trend analysis and advanced data management techniques.

ENVIRONMENT REVIEWS

To maintain constant vigil on environmental compliance, Environmental Reviews are carried out at all operating stations and remedial measures have been taken wherever necessary. As a feedback and follow-up of these Environmental Reviews, a number of retrofit and up-gradation measures have been undertaken at different stations. Such periodic Environmental Reviews and extensive monitoring of the facilities carried out at all stations have helped in compliance with the environmental norms and timely renewal of the Air and Water Consents.

Upgradation and Retrofitting of Pollution Control Systems

In order to keep pace with the changing norms and ensure compliance with statutory requirements in the field of pollution control, NTPC keeps an open mind for Renovation and Modernisation (R & M) and Retrofitting and Upgradation of pollution monitoring and control facilities in its existing stations. It is important to mention that such modifications/retrofit programs not only helped in betterment of environment but also in resource conservation. High efficiency Electro-Static Precipitators (ESPs) of the order of 99.5% and above have been provided at NTPC stations for control of stack particulate emissions. However, the ESPs of a number of stations were built prior to the promulgation of the Environment (Protection) Act, 1986 and notification of emission control standards under this Act. Remedial measures have already been taken up and implemented to improve the efficiency of the existing ESPs at various NTPC stations. ESP performance enhancement programmes by adopting advanced microprocessor based Electrostatic Precipitator Management System (EPMS) was installed at its power stations at Singrauli, Ramagundam, Korba, Farakka, Rihand, Vindhyachal and Unchahar. Additional ESPs were retrofitted in the older power stations, namely at Badarpur and Talcher Thermal. As a result of the above retrofits, the emission of Suspended Particulate Matter (SPM) has been brought down appreciably at the above stations and is maintained within the present statutory limit of 150 mg/Nm3. In new projects, the ESPs have been designed for a maximum permissible outlet dust emission of 50 mg/Nm3 to meet the likely stringent emission norms in the near future.

RESOURCES CONSERVATION

With better awareness and appreciation towards ecology and environment, the organization is continually looking for innovative and cost effective solutions to conserve natural resources and reduce wastes.

Some of the measures include:

- Reduction in land requirements for main plant and ash disposal areas in newer units.
- Capacity addition in old plants, within existing land.
- Reduction in water requirement for main plant and ash disposal areas through recycle and reuse of water.
- Efficient use of Fuel (Coal, Natural gas and Fuel oil)
- Reduction in fuel requirement through more efficient combustion and adoption of state-of-the-art technologies such as super critical boilers

WASTE MANAGEMENT

Various types of wastes such as Municilal or domestic wastes, hazardous wastes, Bio-Medical wastes get generated in power plant areas, plant hospital and the townships of projects. The wastes generated are a number of solid and hazardous wastes like used oils and waste oils, grease, lead acid batteries, other lead bearing wastes (such as garkets etc.), oil and clarifier sludge, used resin, used photochemicals, asbestos packing, e-waste, metal scrap, C&I wastes, electricial scrap, empty cylinders (refillable), paper, rubber products, canteen (bio-degradable) wastes, buidling material wastes, silica gel, glass wool, fused lamps and tubes, fire resistant fluids etc. These wastes fall either under hazardous wastes category or non-hazardous wastes category as per classification given in Government of Indias notification on Hazardous Wastes (Management and Handling) Rules 1989 (as amended on 06.01.2000 & 20.05.2003). Handling and manegement of these wastes in NTPC stations have been discussed below.

Municipal Waste Management

Domestic or municipal waste is generated in households at townships. This waste is seggregated into bio-degradable and nonbiodegradable wastes at source itself in different coloured containers and thereafter the two types are disposed separately. Bio-degradable waste is spread uniformly in identified low lying areas and thereafter it is covered with soil for use later as manure after composting. The seggregated non bio-degradable waste is disposed off separately in other identified low lying areas and is spread out uniformly.

Hazardous Waste Management

NTPC being a proactive organization, the handling and disposal of hazardous wastes are done as per the Hazardous Wastes (Management and Handling) Rules 1989 (as amended in 2003) guidelines issued by Government of India for the treatment, storage and disposal of hazardous wastes. Scientific study on management and handling of hazardous wastes was carried out at a few NTPC stations to adopt the best practices so that there is a complete compliance with statutory requirements. In NTPC sataione, the Hazardous Wastes (Recyclable) are sold/auctioned to registered recyclers/refiners.

The other hazardous wastes such as the activated carbon resins, used drums (hazardous) chromium (Cr-III electrolytes, used petro-chemicals, asbestos packings, used torch batteries, ribbon, toners/cartridges, mixed wastes (waste oil, water and cotton) filters, earth contaminated with synthetic oil (FQF) glass used and sodium silicate, lamps and tubes etc. fall under the category of Hazardous Wastes (Non-Recyalable). These wastes are small in quantity and are stored in properly identified locations. As per the notification, hazardous wastes (non-recyalable) are to be sent to State Pollution Control Board (SPCB) approved common treatment storage and disposal facility (TSDF).

Bio-Medical Waste Management

Hospital (or Bio-medical) wastes get generated from hospitals and they include urine bags, human anatomical wastes, plaster of aris waste, empty plastic bottles of water and glucose, blood and chemical mixed cotton, blood and urines tubes etc. these wastes are segregated and are placed in buckets of different colours as per the notification for Bio-Medical Waste (Management and Handling) Rules.

The seggregated bio-medical wastes are either disposed through the SPCB approved agency or they are treated in autoclaves before disposal into biomedical waste disposal pits. The treated bio-medical waste is spread uniformly and covered with 10 cm thick soil in bio-medical waste disposal pits.

LAND USE/BIO-DIVERSITY

As a policy, NTPC lays special emphasis on land use and Bio-diversity by way of development of green belts, energy plantations, reclmation of abandoned Ash Ponds and EIA and ecological monitoring in the project areas and its surroundings.

Reclamation of Abandoned Ash Ponds

The reclamation of abandoned ash pond sites is a challenging task. NTPC has reclaimed temporary ash disposal areas at some of its projects namely Ramagundam, Talcher Thermal, Rihand, Singrauli and Unchahar through plantation and converted these sites into lush green environments. Extensive plantations have also been undertaken on dry ash mound at NTPC-Dadri. It is planned to reclaim all the abandoned ash disposal areas by plantation.

Green Belts, Afforestation and Energy Plantations

Whats more, in a concerted bid to counter the growing ecological threat, NTPC is undertaking afforestation programmes covering vast areas of land in and around its projects. Appropriate afforestation programmes for plant, township and green belt areas of the project have been implemented at all projects. In order to enhance green cover in the areas around our projects, as a responsible corporate citizen, NTPC till date has planted more than 18 million trees at its projects throughout the country. The afforestation has not only contributed to the aesthetics but also has been serving as a 'sink' for the pollutants released from the station and thereby protecting the quality of ecology and environment in and around the projects. Thrust has also been given to bio-diesel plantation and around 4.8 lakh energy plants including Pongamia and Jatropha have already been planted. A pilot project for extraction of seeds from these bio-diesel plants has also been set up.

USE OF WASTE PRODUCTS AND SERVICES-ASH UTILIZATION

Ash is the main solid waste which is put into use for various products and services. NTPC has adopted user friendly policy guidelines on ash utilisation. In order to motivate entrepreneurs to come forward with ash utilisation schemes, NTPC offers several facilities and incentives. These include free issue of all types of ash viz.

Dry Fly Ash/Pond Ash/Bottom Ash and infrastructure facilities, wherever feasible. Necessary help and assistance is also offered to facilitate procurement of land, supply of electricity etc from Government Authorities. Necessary techno-managerial assistance is given wherever considered necessary.

Besides, NTPC uses only ash based bricks and Fly Ash portland pozzolana cement (FAPPC) in most of its construction activities. Demonstration projects are taken up in areas of Agriculture, Building materials, Mine filling etc. The utilisation of ash and ash based products is progressively increasing as a result of the concrete efforts of these groups.

Advanced/Eco-friendly Technologies

NTPC has gained expertise in operation and management of 200 MW and 500 MW Units installed at different Stations all over the country and is looking ahead for higher capacity Unit sizes with super critical steam parameters for higher efficiencies and for associated environmental gains.

At Sipat, higher capacity Units of size of 660 MW and advanced Steam Generators employing super critical steam parameters have already been implemented as a green field project. Higher efficiency Combined Cycle Gas Power Plants are already under operation at all gas-based power projects in NTPC. Advanced clean coal technologies such as Integrated Gasification Combined Cycle (IGCC) have higher efficiencies of the order of 45% as compared to about 38% for conventional plants. NTPC has initiated a technoeconomic study under USDOE/USAID for setting up a commercial scale demonstration power plant by using IGCC technology.

These plants can use low grade coals and have higher efficiency as compared to conventional plants. With the massive expansion of power generation, there is also growing awareness among all concerned to keep the pollution under control and preserve the health and quality of the natural environment in the vicinity of the power stations. NTPC is committed to provide affordable and sustainable power in increasingly larger quantity. NTPC is conscious of its role in the national endeavour of mitigating energy poverty, heralding economic prosperity and thereby contributing towards India s emergence as a major global economy.

ENVIRONMENT DESTRUCTION

It looks as if environment is an idea whose time has come. Newspapers give prominent display to environmental horror stories. Editorials demand better management of natural resources. Government statements on the need to preserve the environment are commonplace. Government programmes, too, are quite numerous and increasing in number day by day. There are massive schemes for afforestation, for instance. In the last four years, some 1,000 crore seedlings are said to have been distributed or planted. There are new laws for control of air and water pollution and for the conservation of forests.

India has been praised all over the world for what it has done to preserve tigers. Nearly three per cent of India's giant land mass is now protected national parks and wildlife sanctuaries, and there are demands to strengthen their protection and increase their area. Plan documents and party manifestoes take care to mention the importance of environment. But there is a major problem with this entire range of activities and concerns: it does not appear to be based on a holistic understanding of the relationship between environment and the development process in this country.

The programmes are ad hoc, without clear priorities, and there is too much of a policeman's attitude. They seem to be based on the belief that concern for the environment essentially means protecting and conserving it, partly from development programmes but mainly from the people themselves. There is little effort to modify the development process itself in a manner that will bring it into greater harmony with the needs of the people and with the need to maintain ecological balance, while increasing the productivity of our land, water and forest resources. The post-independence political debate in India has centred on two major issues: equity and growth. The environmental concern has added a third dimension: sustainability. India's biggest challenge today is to identify and implement a development process that will lead to greater equity, growth and sustainability. The environment is not just pretty trees and tigers, threatened plants and ecosystems. It is literally the entity on which we all subsist, and on which entire agricultural and industrial development depends. Development can take place at the cost of the environment only upto a point.

Beyond that point, it will be like the foolish person who was trying to cut the very branch on which he was sitting. Development without concern for the environment can only be short-term development. In the long term, it can only be anti-development and can go on only at the cost of enormous human suffering, increased poverty and oppression. India may be rapidly approaching that point. Amongst the hundreds of voluntary groups working at the micro level in India, there has been a remarkably rapid growth of interest in environmental problems.

So rapid has been this growth that sometimes it is loosely described as the beginnings of an environmental movement in the country. Hundreds of field-level groups today take a keen interest in environmental issues and their experiences and interests are extremely diverse. Some are interested in preventing deforestation. Others are interested only in afforestation. Many want to prevent the construction of one dam or another. There are others who want to prevent water pollution. There is the famous Chipko movement in the UP Himalaya, probably the oldest and most famous of all the groups, which has played a major role in bringing the issue of deforestation to the forefront of public opinion.

There is its counterpart in the south, the Appiko movement in the Western Ghats of Karnataka. Dams like the Silent Valley and Bedthi have already been stopped because of strong people's protests, and the well-known social worker Baba Amte is leading a major campaign against the proposed Bhopalpatnam and Inchampalli dams on the borders of MP, Andhra Pradesh and Maharashtra. The Kerala Sastra Sahitya Parishad has had a long acrimonious battle over the pollution of the Chiliyar river in Kerala by a rayon mill. The India Development Service finds itself embroiled in another case of river pollution by a rayon mill in Karnataka.

Meanwhile, the Shahdol Group has worked against the pollution of a river in Shahdol district by a paper mill. There is, also, the Mitti Bachao Abhiyan to organise farmers against water-logging caused by faulty irrigation systems. While all these are relatively well known groups and have attracted varying degrees of media attention, there are many others which are doing excellent work in mobilising people, both to prevent further ecological destruction, often in the face of determined government policies, and to bring about ecological regeneration. One thing, however, that binds most of these groups is their concern to put the environment at the service and the control of the people, the people usually being defined as the local communities who live within that environment.

Environmental protection per se is of least concern to most of these groups, including the well known Chipko movement, for example. Their main concern is about the use of the environment: how should the environment be used and who should use it and benefit from it. It is this growing understanding of the relationship between the people and their environment, born out of a concern for a more equitable and sustainable use of the environment, that is probably the most fascinating development in India today.

ENVIRONMENTAL DESTRUCTION BY THE RICH

To understand the nature of the environmental problems in India, it may be useful to compare and contrast certain environmental trends and concerns in India with those in the West, especially since the environment crusade began in the West and since many groups in India, including political parties, have for long dismissed it as a petty Western concept. The argument has always been that too much concern for the environment can only retard economic and industrial development. The UN Conference on the Human Environment held in Stockholm in 1972 was the landmark conference that created worldwide consciousness about environment.

No UN conference has ever been able to collect so many luminaries at one place. Many delegations from developing countries had argued that the solution to environmental problems lay in economic development. "Smoke is a sign of progress," the Brazilian delegation had thundered, then representing a country witnessing an economic boom. India's Prime Minister Indira Gandhi, who made a major impression on the conference, is still remembered for her oft-quoted statement: "Poverty is the biggest polluter." In all those who came from the Third World, both leftists and rightists, there was a sneaking suspicion that the Western countries were up to some trick.

The West may simply be pushing the environmental concern onto an unsuspecting Third World to retard its technological modernisation and industrial development. It was even argued that having got their riches and their affluent lifestyles, Westerners were now simply asking for more affluence: clean air, clean water and large tracts of nature for enjoyment and recreation, many of which were going to be preserved in the tropical forests and savannas of Asia, Africa and South America. But exactly 10 years later when the UN organised a meeting to commemorate the Stockholm conference, few nongovernmental groups from the Third World were prepared to argue in favour of the development process as it is.

The Third World today faces both an environment crisis and a development crisis, and both these crises seem to be intensifying and interacting to reinforce each other. On one hand, there does not seem to be any end to the problems of inequality, poverty and unemployment, the crucial problems that the development process is meant to solve. On the other, environmental destruction has grown further apace. But what is interesting is that while many environmental problems, expecially those related to air and water pollution have tended to become less severe in many parts of the industrialised world, because of the introduction of highly capital-intensive pollution control technologies, these problems have continued to grow and become critical in many parts of the developing world.

In other words, while the economic development process in the world is only worsening our environmental problems, it is tending to solve them in the West. Michael Heseltine, then minister of environment in the antienvironment British government of Margaret Thatcher, even went so far in a UN meeting in 1982 as to say that all environmental problems in the West have been solved and they now remain mainly in the Third World! Heseltine, indeed, had a point. London, for instance, has not seen for years any of those smogs it saw regularly in the 1940s and'50s, which led to thousands of deaths, and the Thames now even boasts of salmon. But Michael Heseltine, like all representatives of vested interests, was only hiding something of deep significance: the role of the Western world in destroying the Third World's environment.

Very simply speaking, the major environmental problems in the West are those arising out of waste disposal—problems of air and water pollution and of disposal of highly toxic industrial and nuclear wastes. Though problems of acid rain have definitely increased and there does not yet seem to be any solution to the problem of toxic wastes, it is true that some cities and rivers do look cleaner. In the Third World, however, as its own industrialisation proceeds, these waste disposal problems are getting worse day by day but they are still not the major or the only environmental problem.

In the Third World, the major environmental problems are those which arise out of the misuse of the natural resource base: the soils, forests and water resources. These problems are created to a great extent because of the pressure to produce raw materials for modern industry. The Third World's environment not only provides raw materials for its own industries but also for the industries of the West. For instance, the Japanese and Western timber industries have been the biggest source of forest destruction in Southeast Asia. Having turned countries like Thailand from net exporters into net importers of wood, Japanese companies are now turning to the last great wooded frontier of the world: the Amazon basin of South America.

FEEDING THE RICH

The food needs of the Western world have played equal havoc with the lands of the Third World. No statistics on this are available, but if someone did collect them, we will definitely find that despite the worldwide process of decolonisation, there is today many times more land being used in the developing world to meet the food and other biomass needs of the Western countries than in the 1940s, before the process of decolonisation began.

More than a quarter of all central American forests have been destroyed since 1960 for cattle ranching: 85 per cent to 95 per cent of the beef produced as a result has gone to the US while domestic consumption of beef in central America has fallen dramatically. In the US, this beef has been mainly used to make tinned and pet foods and cheap hamburgers because central American beef is half the price of the grass-fed beef produced in the US. The price of central American beef does not represent its correct ecological cost. Cattle ranching has proved to be the worst form of land use for the fragile soils on which these tropical moist forests existed.

Within five to seven years, their productivity has dropped dramatically and cattle ranchers have had to move on. The Sahelian drought of 1968-74 in Africa which grabbed world headlines and claimed the lives of approximately 100,00 nomadic people, was caused by the French colonial policy to drive the region's countries into peanut farming to secure its own source of vegetable oils. Through heavy taxation policies, the French colonial authorities forced west African peasants to grow groundnuts at the expense of subsistence crops. Groundnut cultivation rapidly depleted the soil. It soon spread to traditionally fallow and forest zones and encroached on land previously used for grazing, upsetting the delicate balance between the farmers and nomadic herders.

The expansion of groundnuts was encouraged by artificially high prices but when US soya production began to hit the European market and vegetable oil prices began to fall, the newly independent west African countries had no alternative but to increase area under groundnut to keep up their foreign exchange reserves. As this area increased by leaps and bounds under the pressure of government policies, the nomads were slowly pushed further and further north into the desert, something for which they were not prepared, their traditional relationships with the settled farmers having been totally disturbed. When the long period of drought set in and thousands of animals and human beings began to die, the nomads and their overgrazing was blamed.

Nobody blamed the French or the Sahelian elite which worked hand-inglove with the French. The UN Environment Programmes (UNEP) in a recent report points to the impact of the heavy debt burden of the Third World and high interest rates in the West on the environment of the Third World. The debt burden and regressive terms of trade have forced many developing countries to put enormous pressure on their natural resources, sometimes even to the point of overexploitation.

In 1981, for instance, it took one Latin American country 9.8 times as much beef to buy a barrel of oil as it did in 1981. At the end of the 1970s, profits from the export of one tonne of bananas were enough to purchase only half the steel they would have bought at the end of the 1960s. When interest rates are high, there is a tendency to discount long-term issues like environment for short-term gains. A one per cent increase in interest rates adds approximately US $5 billion to the current debt burden of developing countries. To have increased its export earnings (not profits) by $1 billion in 1981, South America as a whole would have had to increase its banana exports three-fold, Ecuador three-fold and Colombia ninefold, while leading cotton exporters like Egypt and Turkey would have had to double and triple their cotton exports respectively.

This would have meant bringing millions of additional hectares into production to grow these export crops. And, it can be added, this would have pushed millions of marginal peasants into marginal lands like desert fringes and steep hill slopes for their survival, leading to accelerated desertification and soil erosion. In our own country, the first major attack on the forests of the northeast came with the establishment of tea plantations. Destruction of forests goes on for coffee and other export crop plantations even today. The current overfishing on India's coasts, as on the coasts of almost all Southeast Asian countries, is taking place because of the heavy demand for prawns in Western and Japanese markets.

This overfishing is leading to considerable tensions between traditional fisherfolk and trawler owners, and violent encounters between the two are regularly reported. Recently, Indonesia completely banned the operation of trawlers in its coastal waters and several countries, including India, have decreed regulations to prevent trawler operators from fishing in the first few kilometres from the coast.

This zone is reserved for traditional fisherfolk. But policing trawlers over such an extensive coastline is an expensive proposition and regulations are, therefore, seldom observed or enforced. The export of frog's legs to cater to the palates of Western consumers and its impact on the agricultural pest populations in affected areas is now an old story. Fortunately, India not being a major export-oriented country, there are not many such examples. But the pattern of environmental exploitation by industry that we see on the global scale simply reproduces itself on the national scale in India.

What Western industry does to the Third World environment, Indian industry does to the Indian environment. Just to get an idea of how heavily dependent modern industry is on the natural environment, it may be useful to point out that nearly half the industrial output in India is accounted for by industries which can be called biomass-based: that is, industries like cotton textiles, rayon, paper, plywood, rubber, soap, sugar, tobacco, jute, chocolate, food processing and packaging, and so on. Each of these industries exerts an enormous pressure on the country's cultivated and forest lands. They need crop lands, they need forests, and they need energy and irrigation.

The Indian paper industry has ruthlessly destroyed the forests of India. Paper companies in Karnataka, having destroyed all the bamboo forests, are now getting their raw materials from the last major forested frontier of India: the northeast. The government's own public sector paper companies are coming up in the northeast itself. The Andhra Pradesh government has meanwhile set its sights on the forests of Andaman and Nicobar Islands for the paper mill that it wants to build in Kakinada. A leading soap manufacturer has proposed that the Great Nicobar Islands be denuded to plant palm oil. The shortage of raw materials for wood pulp has already forced the government to liberalise import of pulp for the country's paper industry, thus adding to the pressure on the forests of other Third World countries. One lesson is, therefore, clear: the main cause of environmental destruction in the world is the demand for natural resources generated by the consumption of the rich (whether they are rich nations or rich individuals and groups within nations) and because of their gargantuan appetite, it is their wastes mainly that contribute to the global pollution load.

THE ROLE OF SCIENCE AND TECHNOLOGY

If we want to see this process of 'resource colonisation' in a historical light, we can trace it back to the beginnings of the industrial revolution. The cultural diversity that existed in the world at that time was no historical accident. It was born out of the world's biological diversity and people's ability to shape a society with the use of that biological diversity. People came to adopt a lifestyle that was in harmony with the dictates of the immediate environment. Those who lived in the deserts practised nomadism and those who lived in the hills practised shifting cultivation.

Houses and cities built with local materials and skills were built differently to satisfy the conditions set by the local climate. But as modern science and technology grew, it gave people enormous power. Unfortunately the new knowledge created by scientific and technological development fell on an highly unequal world, where access to this knowledge was and still is, unequal, and the infrastructure to use it very uneven. The major result of scientific and technological development, embedded as it is within an unequal global society, is that it has given few people more and more power to exploit resources from further and further away.

First, distant lands (distant ecosystems) were turned into colonies for resource exploitation; when decolonisation took place, multinational corporations took over the role of global resource exploitation and management within a framework of liberal international trade; and now it is the turn of such hitherto unfathomable global commons as the deep sea bed, Antarctica and space to be exploited.

The UN even has a committee to discuss resource exploitation on the moon. But at least on the terrestrial level, with hardly any uninhabited frontiers left, this relentless search for resources from others' ecosystems can only mean extraordinary conflict. In a more and more densely populated world, that which is a resource in a distant ecosystem for one human being is simply a resource within someone else's immediate ecosystem. If that someone's very survival is dependent on the resources of the immediate ecosystem, then there will be conflict, and resource colonisation by the powerful will end in the dispossession and impoverishment of the less powerful, in their own homeland, so to say.

These trends raise serious doubts about the sustainability of the Western industrialisation model based on global management of resources for the consumption of a few. The growth of science and technology has indeed been humankind's most magnificent achievement but definitely not the ends to which this knowledge has been used. The most unfortunate part of this process is that we have now got a large number of consumers in the world whose consumption is highly destructive of the environment but who cannot perceive this destruction psychologically. These are consumers of distant resources: those who have access to all the fruits of modern science and technology. Just imagine a resident of Delhi who uses shirts made of cotton which has been produced in a field in Maharashtra heavily sprayed with pesticides leading to multiple resistance in mosquitoes; electricity from a dam in the Himalaya that has destroyed forests and blocked migration of fish; paper produced in Madhya Pradesh by a factory that has polluted the local river and logged forests in an ever-widening circle, disrupting the life of tribals; cereals from Punjab where food is produced using a technology that drains soil fertility, and so on. It is humanly impossible for such distant consumers to appreciate what their purchasing power is doing to distant areas.

Just imagine, then, the situation of the Western consumer whose beef and fruits come from Latin America and Africa, peanuts from west Africa, coffee from east Africa, tea from India, prawns from Asia and timber from all over the world. But what is probably worse is that such consumers, no longer dependent in any crucial sense on the immediate environment for their consumption needs, slowly become oblivious of the destruction even of the immediate environment. Thus, we get a world economic system in which individuals with power have almost no interest in the fate of their environment, distant or immediate–and only a global environmental crisis, situated amidst a sea of economic poverty, has been able to evoke some concern, and we still talk of the need for environmental education in a big way.

Modern science and technology is not always used as a brutish tool by the rich and powerful to dispossess the poor. It has also been used subtly to undermine the confidence of the poor in their own resources and resourceuse patterns. Particularly, if a natural resource used traditionally by people is so widely available that it just cannot be monopolised, every effort is made in the name of scientific progress to discredit that natural resource as archaic and useless. This drive to discredit a resource is undertaken to the extent that even the poor who often have no alternative except to use that resource, look forward to the day when they can do without it. The classic example is mud. Over half the world's population lives in mud structures. Prices of cement and bricks required for modern buildings continue to rise, being extremely energy intensive, faster than the purchasing power of the world's poor. The chances of all mud dwellers ever moving from their buildings is pretty low, at least not in the next 50 years. The majority of the urban population in the Third World cannot afford even the cheapest modern housing (about US $500- $3,000). In most cities of the developing world, including India, the portion of the population living in slums and squatter settlements has been rising rapidly.

Unable to meet the housing needs of the relatively small urban populations, the governments of the Third World have not even dared to launch housing programmes for rural areas, where the majority of the population lives. The importance of traditional building materials to meet the needs of the world's population is ignored by housing authorities everywhere, and their use is never taught to architects or civil engineers. Mud homes are synonymous with extreme poverty, so much so that they are despised even by the poor and virtually every effort to introduce improved mud housing has failed.

While architects and civil engineers only neglect mud, medical professionals adopt an extremely hypocritical attitude towards herbal medicine. The majority of drugs available on the US market today, for instance, are plant-derived (that is, the drugs were either made directly from a plantbased raw material or they were first discovered in a plant before a chemical company began to market their chemical analogues). Drug companies take great care to screen herbs for pharmacological properties. The moment an useful herb is discovered, the company tries to identify the active ingredient and develop its chemical analogue. This is produced, packaged and marketed and prescribed by the medical profession.

Yet doctors dismiss herbal medicine as outright quackery and make no effort to study medical systems based on herbal medicine. Drug companies make no effort to promote herbs directly. Colonial authorities in Africa had even outlawed the practice of herbal medicine. These administrations had such an effect on the colonised cultures that even after independence, many African countries continued to ban the practice of folk medicine, still the only real option available to a large part of the world's people.

Even more dramatic is the process by which commercial and scientific propaganda has been used to discredit a readily available resource like breast milk, and promote commercial substitutes. Now scientific studies show the superiority of breast milk, but it took a long campaign to formulate appropriate public policies to protect breast milk and even today, vested interests oppose these policies.

THE POOR AND THEIR ENVIRONMENT

The above discussion makes it clear that it is the poor that are affected the most by environmental destruction. The field experience of voluntary groups confirms that eradication of poverty in a country like India is simply not possible without the rational managemet of our environment and that conversely, environmental destruction will only intensify poverty. Environmental destruction goes hand-in-hand with social injustice. A major reason for this is seldom recognised.

The vast majority of the rural households meet their daily household needs through biomass or biomass-related products, which are mostly collected free from the immediate environment. In short, they live within nothing other than a biomass-based, subsistence economy. Food, fuel (firewood, cowdung, crop wastes), fodder, fertilizers (organic manure, forest litter, leaf mulch), building materials (poles, thatch), herbs and clothing are all biomass products. Water is another crucial product for survival.

Water is not biomass itself, but its availability is closely related to the level of biomass available in the surrounding environment. Once the forest disappears, the local pond silts up, the village well dries up, and the perennial stream gets reduced to a seasonal one. The water balance gets totally upset with the destruction of vegetation; in a monsoon climate like ours with highly uneven rainfall over the year, this means greatly increased run-off and floods during the peak water season and greatly increased drought and water scarcity in the lean dry season. There is reason to believe that the number of 'problem villages' from the point of view of availability of drinking water may be increasing. The magnitude of India's dependence on biomass for meeting crucial household needs can be appreciated by looking at the energy situation. We as Indians love to point out that India is the world's 10th largest industrial power, or something like that. But even then, over 50 per cent of the fuel consumption in India is for such a fundamental survival activity as cooking.

In developed countries, cooking accounts for less than 10 per cent of total national fuel consumption. Equally important for India is the fact that 90 per cent of the cooking fuel in India is biomass: that is, firewood, cowdung and crop wastes. Even urban households are heavily dependent on firewood as fuel. Few people, energy planners and government officials alike, had any idea until recently of the dimensions of the rural-urban fuelwood trade.

Annual urban purchases of fuelwood are well over `500 crore in India. Over the last 10 years, Madhya Pradesh, the only state left in the heart of India with any reasonable degree of forest cover, has emerged as major supplier of fuelwood to the cities of northern and western India. But as the state gets more and more deforested every year, fuelwood prices have been rising within Madhya Pradesh itself. Madhya Pradesh has now become India's first state and probably the world's first state too, to ban the export of firewood.

As a result, Delhi, which uses about `15 crore worth of firewood on the retail market, is now getting less firewood, and an increasingly higher proportion from as far away as Assam, the same place from which paper mills are now getting their bamboo. Firewood prices in Delhi are today higher than ever before and according to FAO figures, are among the highest in the world. This firewood is used by the poorest of the poor in Delhi, by construction workers and others. When the effective heat provided by firewood is compared to kerosene, because of the low efficiency chullhas in which firewood is used, firewood becomes four to five times more expensive than kerosene. LPG, incidentally, a fuel used by the richest sections of the society and supplied entirely by government sources, is the most convenient and cheapest fuel available. Which state will become the next to ban firewood exports is anyone's guess. What happens when Assam also says no?

Biomass resources not only meet crucial household needs but also provide a range of raw materials for traditional occupations and crafts and are, hence, a major source of employment: firewood and cowdung are important sources of fuel for potters; bullock carts and catamarans are made from wood; bamboo is a vital raw material for basket weavers, and so on. Traditional craft are not just being threatened by the introduction of modern products but also by the acute shortage of biomass-based raw materials.

A study from the Indian Institute of Science–the first in India on the changing market for bullock carts–reports that people in Ungra village in Karnataka can now longer afford to buy new bullock carts with the traditional wooden wheel because wood has become extremely expensive. A recent report from the Murugappa Chettiar Research Centre from Madras says that traditional fisherfolk now find it very difficult to make new catamarans because the special wood they use is extremely scarce and expensive. Several reports from all over the country–from Madhya Pradesh, from Maharashtra, from Tamil Nadu–portray the extreme difficulty of hundreds of thousands of basket weavers in eking out an existence because of the acute shortage of bamboo.

In the Bhandara and Chandrapur districts of Maharashtra, nearly 70,000 mat and basket weavers have been protesting against the discriminatory prices and small quota of bamboos given to them whereas big paper mills have been leased large bamboo forests. In Karnataka, Madhav Gadgil undertook a study of the use of the state's bamboo forests by paper mills, after a series of protests by basket weavers. Gadgil found that whereas bamboo was available to paper mills at `15 a tonne, it was available to basket weavers and other small bamboo users in the market at `1,200 a tonne. Social activists in Saharanpur have pointed out the travails of the baan makers who have now been deprived of their earlier sources of bhabhar grass. The Uttar Pradesh Forest Development Corporation discriminates in favour of paper mills and this policy has turned thousands of these baan workers into destitutes, landless labourers and urban migrants. Wood is now difficult to get for making even agricultural implements like the plough, especially wood that has been traditionally used for these implements.

Few people know that one of the things that led to the Chipko movement was the anger of the local people over the forest department's refusal to provide ash wood, wood that has been traditionally used for making ploughs, whereas Environment Policy and the forest departments happily allocated the same wood to sports goods manufacturers.

Even biomass resources like thatch have become so scarce that maintenance and repair cycles of mud-and-thatch huts have increased considerably. A government report from Bastar, of all places, as it is still one of the heavily forested districts in the country, points to a village where no new hut has been built over the last two decades because the entire area around the village has been deforested. Traditional mud roofs have almost disappeared from many parts of the country because of the large quantities of timber needed by them.

They are being replaced by tiled roofs, but baking tiles still requires large quantities of firewood. Fodder is another vital resource that is in acute shortage. With only 2.45 per cent of the world's land mass, India supports 15 per cent of its cattle, 52 per cent of its buffaloes and 15 per cent of its goats, and these animals play an extremely important role in the integrated system of agriculture and animal husbandry that Indian farmers practise.

Shortage of fodder, especially from public lands, means, as a study from the tribal areas of Gujarat shows, poor landless households and marginal farmers do not benefit much from the milk cooperatives and animal improvement schemes in the region. In such a situation where millions of people are heavily dependent on biomass sources for their daily existence, the destruction of the environment or any policy that reduces access to biomass resources like the proposed Forest Bill and Forest Policy, will have an extremely adverse impact on the daily lives of the people.

ENVIRONMENT AND WOMEN

The destruction of the environment clearly poses the biggest threat to marginal cultures and occupations like that of tribals, nomads and fisherfolk which have always been heavily dependent on their immediate environment for their survival. But the maximum impact of the destruction of biomass sources is on women. Women in all rural cultures are affected, especially women from poor landless, marginal and small farm families. Seen from the point of view of these women, it can even be argued that all development is ignorant of women's needs at best, and anti-women at worst, literally designed to increase their work burden.

Given the culturally accepted division of labour within the family, the collection of household needs like fuel, fodder and water is left to women. As the environment degrades, and this collection work becomes increasingly difficult, women have to spend an extraordinary amount of time foraging for fuel, fodder and water in addition to household work, agricultural work and caring for animals. There is almost no data which shows how the time spent by women on their daily household activities is increasing and how this increase differs across different eco-climatic zones of India. But the data that is already available on the existing work burden is downright shocking. In many parts of the country, women may have literally reached the end of their 'carrying capacities'. The worst situation is in the arid and semi-arid parts of the country and in the hill and mountain villages. In all these areas, trees and forests have been steadily destroyed. Because of a number of factors–soil and climatic conditions, very small land holdings, lack of irrigation, etc–the Green Revolution has also not reached these areas unlike Punjab and Haryana, where trees are few but where the Revolution has meant an enormous increase in biomass from croplands.

As a result, there is now an acute biomass famine in these areas. In all such areas women can spend as much as five to six hours every day–in some households as much as 10 hours every day–just collecting fuel and fodder. On the contrary, in a state like Kerala, where tropical eco-climatic conditions permit a rich green cover, the work burden of women is much smaller–probably the least in the country. Even the minimal land reforms in which landless families have been given a 10th of an acre, has meant access to a few dozen coconut trees, which helps to provide at least half the fuel requirements.

But in the rest of the country, with deforestation, the increasing work burden on women is affecting everything else in their lives. Women do not even have time to seek health care when they are ill or take rest for few days after a tubectomy operation. A study by the Operations Research Group in western Uttar Pradesh found that even pregnant women work for 14 hours a day. They do this almost till a few hours before the delivery and begin normal work three to four days after the birth.

Just as deforestation affects women most, laws that prevent access to biomass, like the proposed Forest Bill which provides inordinate powers to forest officials, will mean maximum oppression of women: increased demands for bribes and harassment. The penetration of the cash economy is also affecting the relationship between men and women in a peculiar way and is creating a real dichotomy in their respective relationships with nature. Men have become more involved with the cash economy than women.

Women continue to deal with the non-monetised, biomass-based subsistence economy of the household. Even within the same household, we can find cases of men happy to destroy nature to earn cash even though it would create greater hardships for the women in collecting the daily fuel and fodder needs.

The Chipko movement has given us numerous examples of this dichotomy in male-female interests. Even the Chipko experience with afforestation confirms this dichotomy and stresses the role of women in ecological regeneration. Even though many crucial household needs could be met by rehabilitating the local village ecosystem–by planting fuel and fodder trees, for instance–the men do not show much interest in doing so. It is the women who are doing most of the afforestation work organised by the Chipko movement. The new culture created by the penetration of the cash economy has slowly but steadily alienated the men psychologically from their ecosystem. Employment for them means work which can bring in cash. Since employment can be found mainly in the city, there is mass male migration. But even when the men are in a village, a job is still something that earns cash. If such jobs are not available they prefer to remain unemployed and do little to enrich the local ecosystem.

The modern economic system is teaching them that household needs can be met only by earning cash, unlike women, who find that in their daily reality, many vital needs are still collected as a free good. There are few of those caste and class barriers in the Himalayan villages which prevent people from working together as a community. But still women continue to walk miles and miles to fetch fuel, fodder and water every day, while men do little to plant trees in the denuded areas around them. This psychological and economic marginalisation of men greatly helps to expand the availability of cheap labour to the cash economy.

It is not surprising that the eucalyptus-based social forestry, trotted out to be such a great success by the World Bank and the forest departments, is all in the hands of men, all planting trees with the cash motive. Other than employing women as cheap labour in nurseries, these agencies have nothing to show in terms of involvement of women–the very people who deal with fuel and fodder–and the government still calls it social forestry. But maybe this should not be surprising.

Making a fast buck, even at the expense of society and ecology, is probably the most social thing that the World Bank can teach us in a cash economy. Male migration–another major phenomenon in modern society–also seems to be increasing the work burden of women, who then have to take care not only of household needed to collect fuel and fodder grows, agriculture must get neglected. As time for fuel and fodder collection grows and firewood becomes scarcer, the traditional practice of manuring fields will also be given up. Cowdung will be used as fuel, as in the plains, but with the lack of manuring, these fragile hill soils will be exhausted very soon.

It will be a disastrous situation both for the local people and the environment. Already in Ratangiri district, a major source of male workers for the city of Bombay, an enormous amount of cultivable land lies fallow. In this way, women from migrant households suffer a double exploitation: that of their own and of their husbands, who are not given enough of a wage to keep a household together. While the husband sweats for a pittance in the town, the wife keeps the household going in the village by collecting free goods from nature and by working harder on the family farm or as a farm labourer. Because of the increasing incidence and intensity of floods, there has been considerable talk in recent years about integrated watershed management in the Himalaya and in the Ghats. The Himalaya is being described as one of the most threatened ecosystems in the world, which in turn determines the fate of several hundred million people in the Indo-Gangetic plains. But if any action for ecological reconstruction has to be taken in the hills, it cannot be done without the involvement of women. The census data of 1981 shows that all the districts in the country which have high rates of female work participation are situated in the Himalaya or in the Ghats.

In the Himalaya a very high proportion of women workers are also classified as cultivators. Therefore, any programmes which aims at ecological rehabilitation in these areas will have to involve heavily overworked women unless, of course, labour is brought into these areas from outside, which will create tensions of another kind. Fortunately, the experience of the Chipko movement shows that women in these parts, despite their 14 to 16 hour backbreaking work schedule, are extremely keen to participate in such work, especially in tree planting.

Once the women are organised and mobilised, the evidence is that they work with great keenness and fight any obstacles that may be created by men, and we get as a result some of the highest tree survival rates found in afforestation efforts. The biggest ally in the demand for an ecologically and socially sound nature is, therefore, womankind. As exactly similar experiences have been noted in east and west Africa, in Kenya and in the Sahelian countries, there is every reason to believe that this differential interest in nature between men and women is cross-cultural in character.

Male trees and female trees are now common jargon amongst those interested in involving communities in afforestation. All this should not be taken to mean that poor rural households do not have any need for cash. The unfortunate thing is that much of the cash generated by the male does not get spent on household needs. A reasonable proportion of this cash gets spent on products like alcohol and tobacco and artefacts of modernisation like transistor radios.

The situation is creating a new demand on the time of poor women and in some cases, is resulting in a new militancy against nature. Women now have to go out and also earn some cash. Millions of rural women today sell firewood in the towns and cities. If we look at the quantum of firewood consumed in the cities, we can say that at least 2.3 million people must be engaged in headloading–bringing wood on their heads to sell in the towns– making the firewood trade the largest employer in the commercial energy sector of the country. Surveys show that selling firewood is a profession that has grown rapidly in the last 10 to 15 years with the growth in landlessness and joblessness. Most of these headloaders are women, and mostly tribal women. Selling firewood is back-breaking work and brings in extremely little money.

But the women do this because they cannot rely on their men to bring any cash back home. The big advantage with headloading is that this work is generally available round the year. So when no other work is available, headloading at least provides some income. Every headloading woman knows that the forests will be soon destroyed and even this horrible occupation will come to an end.

But they are afraid that if they do not take advantage of the forest now, the foresters will soon sell it off to a contractor. The forest departments are extremely keen to get headloading banned. They have made no study of the phenomenon of headloading. They have not made any long-term plans to meet the firewood needs of the cities. They have obviously not made any effort to connect the two issues.

Foresters have become environmentalists: they say that these women destroy young trees and lop trees excessively: therefore, ban them. No wonder forests are a major issue in most tribal agitations. If a head count is taken at this stage, we will find that the destruction of the environment and its transformation is already affecting on an immediate and a daily basis at least the following groups: artisans, nomads, tribals, fisherfolk and women from almost all landless, marginal and small farm households.

These groups add up to no less than half to threequarters of the country's entire rural population. And unlike the situation in the West, the question of environmental destruction is not an issue related to quality of life but a matter of survival.

TOWARDS HOLISTIC MANAGEMENT

If these are the problems, then what do we do about them? First of all, there must be much more holistic thinking regarding the management of our land and water resources. And this will not be easy unless a determined effort is made. For all the talk about the need for a scientific temper, it must be recognised that the current methodology of scientific analysis carries within itself an extremely unscientific practice: reductionism. It is this reductionist approach that has today produced both natural and social scientists who know more and more about less and less, who know how to cure a disease but create another disease in the process.

Ecology is the first scientific discipline that has actually forced people to integrate and not reduce. This can be illustrated by describing what is happening to the three major components of our lands: forest lands, croplands and grazing lands. The destruction of forests has a major impact on the productivity of our croplands. This happens in two ways. Soil erosion increases manifold and the soil literally gets washed away, leading to an accentuated cycle of floods and droughts. But equally important is the impact of the shortage of firewood on the productivity of croplands. When firewood becomes scarce, people begin to burn cowdung and crop wastes.

In many places, cowdung and crop wastes are now the major sources of cooking energy. Thus, slowly every part of the plant gets used and nothing goes back to the soil. Over a period of time, this nutrient drain affects crop productivity. Add to this the technology of the Green Revolution: the technology of growing high yielding varieties on a limited diet of chemical fertilizers like nitrogen, phosphates and potash. The total biomass production goes up and so does the drain of the nutrients from the soil.

Plants need some 25-odd elements to grow, not just N, P and K. They keep taking these nutrients from the soil and the more intensive the agriculture, without any manure and crop wastes going back to the soil, the faster is the nutrient drain. Today, the district of Ludhiana has the highest yields of many cereals but it also has the highest deficiencies of many micronutrients in its soil. In Punjab, many farmers have routinely begun to use zinc as a fertilizer. Soon it will be sulphur, manganese and iron deficiencies.

The micronutrient fertilizer industry, already a `10 crore industry, is set to become the boom industry of the future. But the net result is that Indian agriculture, because of this and many other factors, most of them related to ecological factors, is showing a consistently declining trend in output when compared to the rapidly rising levels of inputs.

If existing croplands and irrigation water resources are not used well, then faced with a rising population, the demand for colonisation of marginal lands for agriculture will grow. As large parts of the country have excellent soils and enormlus sunlight and the only shortage is of water, government programmes have also promoted the cultivation of marginal lands, especially through the spread of irrigation. Fortunately, the rate of expansion of the cropped area has now come down as compared to the'50s and'60s but enormous ecological damage has already been done.

Even more than forest lands, croplands have expanded on to grazing lands. The result is that graziers have been pushed on to lesser and lesser grazing lands. This has, in turn, led to the overstocking of grazing lands, the destruction of their productivity and the impoverishment of the graziers in the process. The graziers have taken recourse to two strategies in this situation. As the environment becomes more and more hostile, they get rid of the more vulnerable cattle and start keeping goats.

The number of goats in Rajasthan has expanded dramatically–much faster than any other livestock. Environmentalists may howl that the goat is highly destructive of the environment but it is a far better suited animal to the hostile environment that human beings are creating in Rajasthan, Gujarat and Maharashtra. It makes economic sense for the grazier to reduce risk during a period of drought, which is common in these areas. The herders try to solve their problem in yet another way: they begin to use forests as grazing lands. This infuriates foresters who see goats and cattle as the worst evil ever devised.

It is true that India's forests are among the most heavily grazed in the world. But it is not true that forests disappear because of goats and sheep: they disappear only when they are chopped down and this is done usually under the aegis of the forest departments. Animals then move in to stop the regeneration of the logged forest. Animals cannot destroy a closed forest by overgrazing. As forests are disappearing in Rajasthan and Gujarat, nomads in large numbers from these states now enter Madhya Pradesh, still a heavily forested state.

The state's foresters, thereatened by this invasion, banned these 'foreign' invaders by law but the Supreme Court struck down the ban saying that Indians cannot be restricted from going from one state to another. So the interstate movement goes on but clashes are accelerating. Madhya Pradesh's foresters killed 5,000 sheep in one extremely brutal incident on the Madhya Pradesh-Rajasthan border last year and local politicians rushed to rouse passions against the nomads, making every effort to break down the highly synergistic relationship that once prevailed between nomads and settled farmers.

The nomadic animals would earlier bring manure to the fields of settled farmers. Like a walking vacuum cleaner, these animals would gather nutrients from all over and dump them on to the farmers' fields where they would rest and the nomad would even be paid for this service. But today there is open hostility. The Madhya Pradesh forest secretary recently wrote an article in the Times of India saying: "For the love of Mike, please keep these animals out of our forests." If only the animals could be kept out, the forests would regenerate themselves.

But such despair is no answer. Animals cannot obviously be kept out without creating extraordinary conflict. Meanwhile, however, the foresters have found an ingenious solution: plant trees like the eucalyptus which cannot be browsed by animals. Eucalyptus is loved by foresters exactly for this reason. There are many who doubt the ability of eucalyptus to produce more biomass than many indigenous species. But it is indeed ironic that when the country faces an acute fodder crisis, the forester can only plant eucalyptus and produce non-browsable biomass, doing exactly the opposite of what the people need. In fact, eucalyptus is the true weed from the point of view of the landless. It is non-browsable like all fast-spreading weeds and does not benefit the poor unless they own land. But in this manner the cycle of destruction is complete.

The forest departments have destroyed forests by selling off timber to the industrial and urban interests. The firewood shortage and the resulting soil erosion is keeping the productivity of Indian agricultural lands low. Croplands have expanded on to marginal lands and have reduced grazing lands. Animals have moved into forests and are preventing regeneration. Meanwhile, as landlessness and joblessness grow, even groups like the tribals who from times immemorial have lived in total harmony with forests, are turning against forests and want to sell them off as fast as they can.

All the chickens are coming home to roost. Experts sit in grand isolation. Foresters have no interest in fuelwood or in croplands. Agricultural experts have no interest in animals or in grazing lands. Animal husbandry people never tell foresters that they must produce fodder banks. In the life of the proverbial last man–or as I would prefer, the last person of Gandhi–all this isolated thinking brings hovoc.

Life for the poor becomes terrible. Finally, even nature turns against itself. Once denuded, the high solar energy and temperature–the very factors that can bring high productivity and prosperity–begin to bring high desiccation, erosion, destruction and social and ecological poverty. FAO figures show that the effect of unchecked soil erosion on soil productivity is one of the highest in the world, given the ecoclimatic conditions of south and Southeast Asia.

IMPROVING THE GROSS NATURE PRODUCT

Nothing could be more important for planners and politicians today than to rebuild nature. But this can only be done if we re-establish a healthy relationship between the people and their environment. Then only a nature that is useful to the millions, not for making millions, can be re-established.

Regardless of what happens in the West in the next two decades, for all its electronics revolution, its efforts to mine the oceans and its efforts to build solar cells and windmills, and how dramatically this changes the face of the world technostructure, and regardless of how much we may want to catch up with the West in the name of modernisation or out of the sheer compulsions of geopolitics, rebuilding nature and rebuilding its relationship with the people will remain the only way to solve the problem of poverty and possibly even unemployment.

With some 100 millon to 150 million ha of waste and near-wastelands and with a crying need to produce biomass, this country can never get a better opportunity to harness the power of its people to the productivity of its land, to strike at the roots of landlessness, poverty and unemployment, all at the same time. If enough biomass was available, poverty, that is, lack of cash, as defined by economists and by the modern civilization, will not disappear. But the rigours of poverty, the increasing susceptibility to natural emergencies like floods and droughts will be arrested by creating more biomass.

In fact, conventional measurements of poverty based on income data or on food calories are clearly inadequate in a situation where the rest of the biomass needs are becoming increasingly difficult to meet and collecting them on a daily basis constitutes the worst–and growing–drudgery humankind, especially womankind, has ever known. These calculations are not only just inadequate but they also reflect a strong gender bias because they deal mainly with those aspects of poverty–lack of cash–that the male is generally concerned with, but not with those aspects of poverty that the woman deals with–lack of fuel, fodder, water etc.

If we were to construct a concept like gross nature product, we would find that for the poor it is this indicator, which is many times more important than the conventional gross national product (GNP). In fact, we can even say that those who do not get much from the conventional GNP–the poor–are the ones who are most critically dependent on the gross nature product. The GNP cannot be allowed to destroy or transform the gross nature product. Just as economists get very worried about the structure of the GNP, it is equally important, if they have the poor in mind, that they get worried about the structure of the gross nature product.

It is not just the quantity of biomass that is important for meeting basic household needs but also its diversity; sources of biomass within any village ecosystem must be diverse enough to meet the various household needs of fuel, fodder, building materials and herbs, and of raw materials for artisans. The diversity in nature has also acted as insurance during periods of emergency by reducing societal vulnerability. During periods of drought and resulting crop failures, which are recurring phenomena in many parts of India, fruits from trees and roots, leaves and wild animals in the forests used to become an important, alternative source of nutrition. In 1983, the tribals of Chotanagpur survived a drought not because of government assistance but despite government callousness. It is the forests which gave them their nutrition. Surviving on the forest during a drought is common in Bastar.

Wild animals caught in the jungles become a significant source of food during periods of crop failure. A study from Africa found that in times of drought, traditional societies had nearly 150 responses. They even fed the thatch of their huts to the goats. But in a modern village there were only two responses: pray to God (which even the Tamil Nadu government recommended during the Madras water crisis) or migrate to the towns. The combination of trees, grasses, crops, animals and ponds, which we found in almost every village was an extraordinarily interactive and resilient system for emergencies.

Instead of destroying this complex and interrelated system, science must be used to build on it. In other words, it is not enough to preserve biological diversity in just those areas of our country where the flora and fauna are genetically rich and diverse by setting up biosphere reserves and national parks, but that biological diversity must be preserved and/or recreated in every village ecosystem. Concentrating on the production of a few commodities (cereals, for instance) is totally inadequate in a society which is only partly monetarised and where the vast majority still has to depend on access to free biomass resources from the immediate environment.

Every village has to become a biosphere reserve. If suff-icient biomass could be generated, there would also be none of those tensions we see around nature protection areas today. At a time when biomass is in acute shortage, it is obviously foolhardy to attempt such protection without massive schemes for largescale biomass generation. Understanding of the gross nature product and how it is changing within the national ecological space (or shall we say, national economy) is extremely limited, despite its crucial importance for the poor. We know nothing, for instance, about the importance of so-called weeds for the poor. The reason we know so little about the gross nature product is because the 'growing stock' of biomass does not get reflected in GNP calculations anywhere.

Therefore, whether the 'growing stock' lives or dies, exists or disappears, does not make any difference to economists. Most economists, not surprisingly, have no clue of what hapens to a subsistence economy when its biomass resources are affected. The increasing hardships and tensions only get reflected in studies of 'political economy which record growing conflicts over land, forests, fodder, grazing lands, water sources, etc, or in totally unexpected indicators like the sex ratio of the country. Just why 22 million women failed to show up in India's 1981 census is anybody's guess.

Surely their extraordinary work burden, in a situation of malnutrition, must be an important factor. It is for this reason that an indicator like gross nature product, and changes in that indicator, would probably reflect far better the changing reality in the subsistence sector of a country like India. Unfortunately, we do not know as yet how to construct such an indicator. But if we did, we will definitely find that while the conventional GNP has gone up, the gross nature product has steadily gone down, the former acting as a parasite on the latter.

The answer to India's immediate problem of poverty lies in increasing the biomass available in nature in a manner that access to it is ensured on an equitable basis. But giving a relevant 'green cover' to the country—the real Green Revolution—would probably require the most holistic thinking that planners economists and scientists have ever known.

The conflicts and complementarities in the existing land use patterns have to be extremely well understood. Otherwise, landuse patterns will remain as chaotic as today. Landless and poorer peasants will continue to oppose planting trees on comunity lands under so-called social forestry programmes because they are afraid this will take away their grazing lands, while forest departments and richer peasants will only plant those trees which animals cannot touch (like eucalyptus) even though there is a stark fodder crisis all around.

Nothing could take us closer to Gandhiji's concept of gram swarajya than striving to create village ecosystems which are biologically diverse and selfreliant in their local biomass needs to the maximum extent possible. This will clearly demand an extremely intensive use of our natural resources like land and water to create huge and diverse growing stock of biomass. Any science which teaches how to do this will truly have the right to be called a people's science—and indeed it will have to begin with the traditional knowledge of the people.

Even more so, planning for the enhancement of village ecosystems will call for village-level planning with the involvement of the people–a level of decentralisation that has never been attempted either in resource planning or in resource management. The biggest challenge, therefore, is before social workers and politicians who have to play a crucial role in ensuring that people can participate in this biomass-based development process. No biomass-based strategy can succeed without the involvement of the people, especially women.

If useful nature is to grow, it will have to be protected and nurtured. The walls built to protect nature will be respected by the people only if they are people's walls and if the people know that what grows beyond, belongs to them. Easy availability of biomass leading to a reduction in women's work burden could create the appropriate conditions for many desirable social changes. Kerala, for instance, is often cited as an interesting case of an economically poor state moving ahead with its demographic transition. But is Kerala poor in environmental terms also?

The availability of biomass, in fact, appears to be the maximum in India and women's work burden low. This may sound an exaggerated proposition but would it be right to surmise that, among many other things, it was also the easy availability of biomass and relatively lower work burden that created the appropriate conditions for literacy programmes for women to succeed? If women in Kerala had to spend as much time working as women do in large parts of Rajasthan, Uttar Pradesh and Bihar, wouldn't there have been immense male pressure to keep women working in crucial household activities and away from schools?

Immediately, at least, the country must realise that a clear biomass policy is desperately needed, which recognises the competing uses for biomass in society, especially between biomass-based industry and poor households and sets clear priorities on the use of biomass in a situation of scarcity. The needs of the poor must be specified as a priority use of biomass in the existing situation of environmental degradation. A beginning definitely needs to be made with the proposed Forest Policy and with rural energy planning in general.

If India fails to recreate nature on a massive scale in a manner that generates employment and equity, not only its villages but also its cities will become unlivable. Many people prefer to call the urban migrants economic refugees from the countryside. But to my mind many of them are really ecological refugees, displaced by dams, by mines, by deforestation, by destruction of grazing lands, by floods, by droughts, by urban expansion, and what not. We have today the world's fourth largest urban population. Before the end of the century we will be the largest. Managing this huge urban population will call for extraordinary political and managerial sagacity and altogether new approaches, something we cannot learn from the rest of the world. But one thing is definite–if the process of urbanisation continues to create the same demands on our rural environment, it will only accelerate the destruction of that environment and in turn make the urban environment impossible to manage.

India cannot survive without a low-energy, low-resource input urbanisation. In its absence, no law or laws, like the so-called Delhi Bills, which try to turn the incoming ecological and economic refugees into our cities into criminals, will work. Only a holistic approach to our problems and dedicated political will to solve them will.

❐

10

Atmosphere

TROPOSPHERE AND STRATOSPHERE

At about the close of the last century, soundings of the atmosphere were begun with the aid of small balloons carrying light devices that automatically registered the temperature and pressure throughout both the ascent and the descent—a kind of exploration that soon was taken up at various places. From these records, in turn, the heights corresponding to given points on the traces were readily computed. In this way we gradually have come to know the average temperature and pressure of the air at every height from the surface of the earth up to at least 12 to 15 miles, under different weather conditions, for all the seasons, and in many parts of the world.

Pretty soon a means of registering the humidity was further added to the apparatus carried, so that we now have a fair knowledge also of the average vertical distribution of water vapour that corresponds to each particular type of weather. Obviously, too, these sounding balloons, as they are called, afford some knowledge of the direction and velocity of the wind at various levels for the particular time and place at which a flight is observed.

From the data thus obtained several interesting generalizations soon became evident. The most conspicuous of all, and for a long while the most doubted, because it was not understood, was the fact that at 6 or 7 miles above sea level, in middle latitudes, and generally 8 to 10 in tropical regions, the temperature no longer decreases with increase of height, but remains substantially constant.

How far this equal temperature, or isothermal condition, extends is, of course, unknown, but it does go at least to the greatest altitudes yet attained, that is, 15 to 20 miles. It may extend to the limit of the atmosphere of appreciable density, or it may not. We have no direct and positive evidence of either alternative. This much we know.

From the surface up to a considerable distance above sea level, generally 6 to 10 miles, depending mainly on the latitude, the temperature decreases at the average rate of about 1°Fahrenheit in 300 feet; then almost abruptly, as a rule, practically ceases to change with further ascent. Clearly, therefore, this unsuspected and striking phenomenon divides the atmosphere into two great parts; a lower portion in which temperature rapidly decreases with increase of height, and an upper in which the temperature is nearly independent of height. In the lower, convection, or ascent and descent of the air, can occur under certain conditions, because although ascending air cools by expansion its ascent brings it into air that also is colder.

Since convection here is impossible masses of air forced into this region would spread out in horizontal strata, and indeed there commonly are evidences of the existence of just such strata, though we seldom if ever are sure of their origin. This filler structure suggested the other, and now the all but exclusive, name, "stratosphere," of this region.

Thus the troposphere and stratosphere are quite distinct from each other. Clouds and every sort of precipitation, rain, snow, graupel, sleet and hail, and every kind of storm, are forever agitating the former, but never for a moment disturb the serenity or overcome the stability of the latter. They are sharply separated, the one from the other, along an approximately horizontal but invisible surface called the tropopause, and so called because that is where convection ceases, as above explained. Perhaps it will be interesting to recall here that sometimes we use the expression, "the sphere," to mean the earth as a whole; that in our first approach to particulars we divide the sphere into the lithosphere, or solid portion of the earth, the hydrosphere, or water portion, and the atmosphere; and that the atmosphere in turn consists of two great concentric shells, the troposphere and the stratosphere. The lithosphere and the hydrosphere also have interesting structures about which many a fascinating tale has been told, but this is not the place to repeat them; ours is the story of the atmosphere and of it alone. When we examine the troposphere closely we find that it too has structure.

Some of these parts are fleeting, but others are at least semi-permanent. One of the more nearly constant is the trade wind, or, better, the trade winds, as there are several winds properly so designated. Advantage has been taken of them, of course, in shipping, especially by sailing vessels, but this is not the source of their name. It does not derive from any idea of commerce, but from the fact that their characteristic is persistent blowing along a particular trade, that is, tread, track or way. The trades are east winds (winds from easterly points) over the tropical oceans, or, more exactly, between the latitudes 30°N and 30°S. There really are five such winds that are well defined, one over the north Atlantic Ocean between latitude 30°N, roughly, and the Atlantic doldrums, or region of calms near the equator, and another over the south Atlantic between latitude 30°S, also roughly, and the same doldrums. Two other trade winds are similarly located over the Pacific Ocean, and there is one over the Indian Ocean, south of the equator, between Australia and Madagascar.

These wind currents, the trades, are very shallow near their poleward boundaries or edges, but grow gradually deeper and deeper to a maximum of at least 4 or 5 miles as their equatorial borders along the doldrums are approached. They are well defined and important structures of the troposphere.

Antitrades

Immediately over the trade winds are the antitrades, or winds that blow away from the equator and at the same time turn more and more nearly eastward and come down lower and lower with increase of latitude until they merge with, and become a part of, the prevailing westerlies-the winds beyond latitude 30°N or 30°S, as the case may be, that from the surface up to at least well into the stratosphere usually are from westerly points. The antitrade (there is one over each trade wind) represents a part of the return to higher latitudes of its accompanying trade, which through most of its route approaches closer and closer to the equator.

The rest of the trade, joined by a greater or less amount of tropical air, turns around the western end of a high pressure ridge, or ocean area of prevailing light and variable winds and clear skies. The frequent and deep southerly winds onto the United States from the Gulf of Mexico and the Atlantic Ocean are of one of those great joint trade and tropical currents.

And the troposphere has many other parts. Two of the strangest and most important of these, strange because though apparently impossible companions they yet are always together, are the cold wind and the warm wind that peacefully flow beside each other—that, in technical terms, are each in dynamical equilibrium with the other. We know that a column of warm air will not stand up in equilibrium with an adjacent column of cold air, and we know that we know it by the way cold out-doors air pushes up and away the warm air in a heated chimney or smoke stack.

Hence we are unwilling at first to believe that a cold wind can blow alongside of a warm wind and not drive it away that the twain can meet on equal terms and each hold its own. But they can and do, because an object, such as a mass of air, whenever moving over the earth tends always, because of the earth's rotation, to turn to one side or the other (the right, going with the wind, in the northern hemisphere, the left in the southern) of the course it is on. And the magnitude of this tendency (the force the moving object would exert on a frictionless vertical surface that would hold it to a fixed geographic direction) is proportional both to its mass and to its velocity. In the case, therefore, of adjacent stationary columns of warm and cold air only gravity is operative and the heavier (because colder) column underruns and buoys up the lighter. This is why air rushes up heated chimneys and smoke stacks why they draw, and the higher the better.

In the case of the winds, however, the situation is quite different, for in proportion to their mass and their velocity they now exert (or would, against a suitable restraining wall) a horizontal or deflective force that also must be considered. If, then, the winds were in the right direction, and had the proper positions with reference to each other, the colder from the east, say, and the warmer just south thereof and from the west, or, more generally, if they were passing each other counter clockwise (in the northern hemisphere, clockwise in the southern) they might have such velocities that the resulting deflective force would just balance the difference in pressure due to difference in density.

Cold Front

When the breakdown between the warm and the cold currents is well developed, with the mass of cold and, of course, polar air pushing its way equatorwards while the warmer air flows beside it poleward, we speak of the interface between them, or, rather, of the intersection of that interface with the surface of the earth, as a cold front. We also call this locus a windshift line because here the direction of the wind changes, as one system, the warm winds, departs and another, the cold, comes on. It likewise is called a squall line because all along it the winds are turbulent and squally and, in summer, often accompanied by thunder storms. This cold front, then, is the boundary, along the surface, between distinct wind systems between a system of relatively cold and dry winds under skies and from higher latitudes, and a system of comparatively warm and humid winds under clouded skies and from lower latitudes two more of the great parts in the structure of the troposphere.

Warm Front

The warmer of the two passing winds just mentioned soon reaches, in the course of its poleward travel, the colder air of higher latitudes. But here, as elsewhere, two masses of air, having different temperatures can not be in equilibrium with each other at rest and standing side by side. They must be, and are, in motion, and in such manner as to retain for a time, and to a greater or less extent, their individuality.

Here again the colder air is wedged under the warm which flows not along this wedge, but up its slope slantingly as a rule, but up, nevertheless. The line along which this part of the interface between the cold and the warm winds cuts the surface is called the warm front.

Thus the cold and the warm sectors of the traveling cyclone maintain their independence as polar winds and tropical airs, respectively, until gradually separated from their source, after which they soon are brought into like conditions each with the air of its new location and merged with it. In this way polar air becomes tropical air, and tropical winds polar winds, back and forth ceaselessly and indefinitely; but always as entities in their travels as elements in the structure of the stratosphere just as rivers and lakes are entities in the ceaseless round of water from oceans to continents and continents to oceans.

Inversions

Normally the temperature of the lower atmosphere, or troposphere, decreases at every level with increase of height, but there are exceptions to this rule; and, besides, the rate of increase always is more or less irregular. A very common exception occurs next to the surface. Indeed, during still clear nights this exception is itself the rule, and practically without exception. At such times, owing to the rapid net loss of heat by the surface through radiation, the adjacent air, by contact with the cold surface, becomes cooled to a distinctly lower temperature than the air at a slightly higher level.

The height to which this cooling extends depends upon the amount of air movement and consequent turbulence or mixing. If this is slight the cooling is restricted to near the surface, but is all the more pronounced. If the movement of the air is appreciable the cooling is less pronounced but extends higher. Finally if there is considerable wind the mixing so distributes the loss of heat as to be small at any level, even at the surface. When, however, the cooling is marked and restricted to a shallow layer the temperature increases through this layer with increase of height instead of decreasing, as is the rule, from the ground up. This is an inversion of the usual temperature gradient, or a temperature inversion, or, for short, just an "inversion."

For a while the air next above a surface inversion layer can slowly blow along over it without rapidly mixing with the colder air or carrying it away thus providing another case of structure in the troposphere. Appreciable winds, however, do wear away such a layer rather rapidly. They also commonly tend to produce, and frequently do produce, a temperature inversion some distance, perhaps a thousand feet or more, above the ground. The surface inversion, discussed above, occurs, as explained, on still clear nights and is due to loss of heat by radiation, and therefore might be called a radiation inversion. The inversion now under consideration may occcur any time the wind blows, because it is due to the turbulence or vigourous mixing of the lower air incident to surface friction and the interference to free flow by trees, house and other irregularities.

Turbulence produces a temperature inversion the turbulence inversion in this way: Before the wind sets in, the temperature of the lower thousand feet or more of the air may and often does, decrease slowly with increase of height, perhaps only one degree

Fahrenheit in 400 to 500 feet. It even may increase with height, especially of early mornings, through the lowest levels, as just explained.

In either case a complete mixing of the air from the surface up to the level of 1000 feet, says, brings all the air, not to a common temperature, as the stirring of water smooths out any thermal inequalities it may have had, but to such temperature that wherever, within that layer, a portion of its air may be taken, up, down or sidewise, it will, on arrival, have precisely the same temperature as the then surrounding air at the same level. That is, it will come to such temperature that nowhere will a moving portion of its either give heat to, or take heat from, the air with which it at any time is in contact. Obviously, because if the agitated air is not yet in a state of temperature equilibrium with a moving portion of itself, all that need be done to make it so is to stir it up further and mix it more thoroughly.

Now an isolated mass of air that neither gains heat from nor loses heat to the atmosphere through which it passes, obviously must cool with ascent, owing to its loss of heat incident to its expansion against the decreasing pressure. Where the air is unsaturated, so that no fog or cloud is formed, and thoroughly stirred up, as above explained, the rate of this decrease, and therefore the rate of temperature decrease in and of the layer itself is, approximately, 1°F per 190 feet, allowing for the average amount of humidity.

This, as explained, is a much faster rate of decrease of temperature than usually exists in the lower air. Hence the mixing of the lower air by turbulence so redistributes its heat as to make the lower portion warmer than it was, or otherwise would be, and the upper portion colder. Immediately above the topmost portion of the turbulent layer the air is undisturbed by vertical convection and therefore distinctly warmer than the upper portion of the agitated stratum. The turbulence inversion (inversion caused by turbulence) is a sharp partition between two portions of the troposphere, one the lower, full of irregularities, the other smooth-flowing.

It is interesting that as the temperature changes almost abruptly, perhaps several degrees, with change of height at the level of this turbulence inversion, so also must the density of the air change abruptly. And, furthermore, if the pressure gradient or push that causes the winds, is practically the same, as it seems to be, on either side of, and close to, this level, then there also must be here a nearly sudden jump in the wind velocity, from slower in the under, denser air to faster in the lighter air immediately above. However, this change of velocity, seldom more than one per cent of the whole, always is too small to be of any practical importance.

OVERFLOW STRATA

As explained above, any layer of air that is thoroughly mixed up has a certain rate of decrease of temperature with increase of height, and such that the temperature of an isolated mass of like air rising or falling through it will change at the same rate. In such a layer vertical convection is as easy as horizontal gliding over a smooth surface.

Left to itself, though, and if unclouded, its vertical temperature gradient or lapse rate (lapse, for short) gradually changes, owing largely to gain and loss of heat by radiation, until it becomes decidedly less, and the layer thereby impenetrable to dry or unsaturated air. Saturated air, however, may be buoyed up to considerable heights, as already explained, because the heat of condensation, or heat rendered sensible as a result of condensation, so reduces the lapse rate that the ascending air is warmer and therefore lighter than the air surrounding it. The extent to which this rising air is warmer and lighter than the air through which it is passing depends, of course, on the amount of condensation. This in turn depends on the amount of water vapour present, and that depends on the temperature.

Hence, in general, and starting from the same level, saturated warm air is pushed up to greater heights in the process of convection than is saturated cold air.

In any case, though, the amount of condensation per given increase of height, and therefore the quantity of heat available for further convection, becomes less and less with gain of altitude, and finally, at one level or another, insufficient to induce further ascent. At this level then, whatever it is, the rising air spreads out in a sheet or stratum that differs in humidity, temperature and lapse rate from the atmosphere of every other level, both higher and lower. In this way, that is, from convections, great and small, including the over- and underrunning associated with general or cyclonic storms, the troposphere is largely built up of overflow strata.

They are not, of course, the same from day to day nor from place to place, but everywhere they are always more or less distinct and numerous. They often mark the levels of cloud layers of the sheet or stratus forms, of altocumuli and of the windrow or billow clouds, due to the waves caused by the flow of one stratum over another, much as water waves are induced by wind. Occasionally, two adjacent strata differ from each other so radically that a balloon can float a long distance with the bag in the one and the basket in the other.

Of course the identity of each particular stratum ultimately is lost through mixing with others above and below it, whether caused by the vigourous stirring incident to a general storm, or by virtue of the perpetual diffusion and prevalent turbulence over every interface. But so long as it does exist it may be pushed up bodily to greater heights by underrunning air, or depressed to lower levels by an overflow current, with, in either case, a change in the temperature gradient or lapse rate (except in the very unusual case when it initially is that of completely stirred-up air) and a corresponding change in its stability and resistance to penetration by convection in either direction, upward or downward. If the layer is pushed down, without lateral contraction or expansion, by an overflow of air above it the pressure on it will be increased by the same amount throughout, but it will be compressed most, and thereby heated most, on the upper side where the initial pressure is least, and compressed and heated least on the under side where the initial pressure is greatest.

This changes the temperature gradient in the stratum so depressed, except rarely, as above explained, and in such manner as to render the layer increas ingly difficult of penetration a firmer floor and a more rigid ceiling. On the other hand, if the stratum is lifted to a higher level it is cooled most on top and least at the bottom and its effectiveness as a barrier to convection corresponding decreased.

It should be noted that in atmospheric convection, and the consequent production of air strata, water vapour plays a most important role. With increase of humidity, under constant temperature and pressure, the density of the air steadily decreases, just as it would with increase of temperature at constant pressure. This fact probably accounts for many small waterspouts starting at the surface starting there because the lower air becomes relatively light through high humidity, analogous to the starting of dust whirls over a dry region due to decrease of density incident to increase of temperature. There is, however, a fundamental difference between the surface waterspout and the dust whirl.

The latter consists of dry air made light by increase of temperature, and can ascend (be pushed up) until it has lost a certain amount of its original heat and no further. Not so with saturated or highly humid air. It, too, like the dry air, ascends because it is lighter than the adjacent air around about, but it maintains this relative lightness a much longer time through the latent heat rendered sensible by progressive condensation, and thereby reaches far higher levels.

In respect to convection dry air and humid air are like unto two men in business, one with a working capital but no reserve assets; the other having, in addition to his ordinary current needs a much greater fund that may be drawn upon whenever required. The one, like dry air, may start well, but his power to expand soon is exhausted. The other, by drawing on his reserve, can take advantage of every opportunity and thus rise to a far higher level of success. Just as it takes money to rise high in the business world, so too it requires water vapour to make any considerable ascent in the atmosphere.

The konisphere (dust sphere) and its layers. Not all we breathe is air. With every breath we inhale a million microsticks and -stones and a host of other things that are no part of a pure atmosphere. "Where do they come from?" The heavens above and the earth beneath. Every wind that sweeps a desert catches up tons, and sometimes millions of tons, of pulverized rock to spread far and wide. Fragments of vegetable fibre litter the soil the world over and are wafted hither and yon as even the gentlest breeze may blow.

Pollen of conifers, ragweeds, and a thousand other trees and plants we must take into our lungs from spring to fall every day we breathe the open air. And our bronchial tubes need chimney sweeps (luckily provided by Nature) to get rid of their coatings of soot from kitchens, factories and forest fires. Even the ocean, through its evaporated spray, makes a salt mine of the air that we breathe. Then, too, lightning sprays nitrogen acids into the atmosphere, while soft coal and volcanic vents similarly add the sulphur acids but all are too dilute really to bother us. Spores and microbes of many kinds we just have to inhale, for they are everywhere.

And as if all this were not enough the earth, every now and then, explodes at some great volcano and hurls tons upon tons of rock powder into the air where it drifts far away for weeks, months or years, according to its degree of fineness and initial height attained. Finally, in addition to all this dust of its own the world stirs up, the atmosphere to its outermost limits is filled with the ashes, so to speak, of daily millions of incinerated meteors, or shooting stars.

That is how the earth got its konisphere (dust shell). If it had no atmosphere it would have no konisphere, but having an atmosphere it must also have a coexistent and coextensive konisphere.But this konisphere is not uniform; it has distinct layers that, like other phenomena, show structure in the atmosphere. The more pronounced of these layers are:

- The turbulence layer, that is, the layer of air next to the earth that, owing to surface friction, any appreciable wind fills with turbulence. Incident to this churning up of the air there also is a stirring up of the dust. At such times this is thedustiest portion of all the atmosphere, and it carries the largest particles. Its depth is, of course, that of the turbulence, and therefore may be anything from two or three hundred feet up to two or three thousand; while the amount of dust, as determined by the strength of the wind and condition of the surface, can vary from practically nothing at all, as over snow fields, to that of the terror of the desert—the blinding and stifling sand storm. Its upper boundary is rather sharply marked and often distinctly visible from any higher level. This layer includes also the city pall, that 4 tons a day, per square mile (average for Chicago, and there are worse places) smudge of soot and dirt that shuts out so much of the health-giving radiation of the sun.

- The convection layer, or stratum of diurnal convection, marked by the dust carried up from near the surface by warm ascending currents. It therefore is deepest and dustiest during summer droughts, and over arid regions. Its upper surface, perhaps two miles high, frequently is seen by the aviator or balloonist almost as distinctly as the surface of an ocean, and even to resemble that surface through the emergence above it here and there of cumulus clouds that look like so many islands. This is the next dustiest of the shells of the konisphere, but even so its burden seldom is heavy enough to bother in any way those who move about in its densest portion—at the surface of the earth.
- The tropic layer, or layer coincident with the troposphere, and therefore 6 to 7 miles deep in middle latitudes and two or three miles deeper, on the average, in tropical regions. Its top is the limit of even occasional convection, and the dust of its upper levels relatively both sparse and fine. Its upper surface rarely has been observed, since aeroplanes and balloons seldom pass that level. We know where that surface is, however, because every ascending current of air necessarily carries with it some of the dust of the lower levels, and therefore dust of terrestrial origin must extend to the upper limit of convection, that is, to the top of the troposphere, and no farther. We also have observational evidence of this upper surface through the effect of the dust particles on sunlight.
- The stratic layer, or the region of all the atmosphere of appreciable density beyond the troposphere. The dust of this region is of two parts; one roughly constant in amount, the other extremely variable. The first comes from the myriads of meteors that hourly enter the atmosphere. The second is due to occasional volcanic explosions of great violence which, like those of Asama in 1783, Krakatoa in 1883, Katmai in 1912, and many others, hurl powdered rock far beyond the levels of the highest clouds.

The heavier dust particles of the turbulence layer quickly settle of their own weight. To a less extent that is true also of the dust in the convection layer. In the main, however, the finest particles in the troposphere are carried down by condensation, either of water vapour directly onto them or as a result of being picked up by falling drops or drifting snow flakes. In this way the whole lower atmosphere from the surface of the earth to the tops of the highest clouds—a layer 6 to 10 miles thick—is literally washed, or scrubbed, as such processes are called, by rains and snows. If the air were so dry that there could be no precipitation it quickly would become suffocatingly filled with fine dust. In fact, it is believed by some that our sister planet, Venus, has just such a waterless, dust-filled atmosphere. The dust of the stratosphere is not so fortunate, if we may put it that way, as that of the lower levels. It must get down through this region by itself, for there is not enough vapour up there to lend it any aid on its earthward journey. Often it is years in getting out of this arid realm, but once it has covered that part of its course the rest of the trip is quick and easy by way of the snowflake and raindrop routes. Even the dust of the earth, therefore, reveals a considerable structure of the atmosphere—at least four distinct layers. And it falls into still other great divisions according to this or that basis of separation. Nearly all the foregoing concerns the troposphere. The little that follows relates to the stratosphere, about which our knowledge still is very slight.

Upper Trades

Since the stratosphere is much warmer, 30°F to 40°F, in the polar regions, than in the equatorial, it would seem that there must be an upper interzonal circulation of the atmosphere somewhat like a mirrored image of the lower— towards the poles in its under portion and from them in its upper levels. The rotation of the earth obviously would affect this upper interzonal circulation in the same manner that it does that of the troposphere, and therefore lead to east winds over the tropical and adjacent regions, and west winds over the higher latitudes. Furthermore, a little calculation based on the temperature of the atmosphere at various levels in high and low latitudes shows that this upper circulation must begin at the height of 10 to 12 miles.

This calculation further shows that at that level the average pressure must be nearly the same everywhere, and therefore the average wind at this level very light. Both these conclusions, namely, that the winds 10 to 12 miles above the surface of the earth must be light, and above that level from the east in tropical and adjacent regions, are supported by all the observational data (a fair amount) we have on the subject.

Twilight Top

A small beam of sunshine in a darkened room because there are dust particles in the air that scatter the light. That is why a few whisks over the floor with a dry broom makes the beam brighter. This explains, too, why we can see the shaft from the searchlight, and the focusing "streams" when the sun is "drawing water."

In each case the contrast is between the myriads of illuminated motes and the shaded, hence darker, portions of the surrounding air. Not only the dust particles, but also, though to a far less degree, all the gas molecules of the atmosphere, are luminous in sunshine. This air luminosity is the chief factor in the blue of the sky, and an important factor in other sky colours. It accounts also for the twilight arch the visible boundary between the shadow of the earth and the illuminated atmosphere that rises above the eastern horizon as the sun sinks beneath the western.

Since the observer is himself within the earth's shadow it is obvious that by noting the exact time this arch is directly over head, say, he may know, from certain astronomical tables, just how many degrees the sun is then below the horizon; and that from this value, in turn, and the radius of the earth he can compute the height of this arch, that is, the greatest height at which the density of the air still is sufficient to scatter a perceptible amount of incident sunshine.

Numerous measurements of this kind have been made, and that height thus found to be about 44 miles. In respect, then, to its efficiency as a lightscattering agent also the atmosphere has structure, an inner shell about 44 miles thick in which the scattering is appreciable, and an outer in which it is imperceptible.

Auroral Base

The polar lights, both northern (aurora borealis) and southern (aurora australis) divide the atmosphere into distinct parts, an inner, about 62 miles deep, into which auroras generally do not penetrate, and an outer of unknown thickness, but certainly hundreds of miles, in which they commonly do occur.

Kennelly Heaviside Layer

When radio-telegraphy over long distances was first attained we were much puzzled to know how it could be, for surely radio waves are just greatly magnified light waves, and light doesn't bend to the curvature of the earth in such manner that an object can be seen a thousand miles away. But if the air were highly transparent, and both the earth and the encircling sky excellent reflectors, then a powerful light at London say, might well be seen from New York, or any other place on the globe.

The light could not get through either reflector and therefore would keep on traveling between them until finally absorbed. The same is true also of electric waves, and for these the earth is a reflector. If therefore, the sky reflected them too we would expect long distance radio communication to be possible, but not otherwise. Hence, when such communication was accomplished, Heaviside and Kennelly told us that the sky must be a reflector of electric waves, that is, an electric conductor, and everybody answered. "Why, of course, it is a conductor." And then came the long search to find how it is made a conductor and at what level its conductivity is adequate to account for the phenomena observed.

We believe now that this conductivity is owing essentially to the presence in the upper atmosphere, 30 to 60 miles or more above the surface, of a million or so free electrons per cubic inch, due to solar radiation in the far ultra violet. The height of the under surface of this reflecting region, or Kennelly-Heaviside layer, as it commonly is called, is not sharply determined since it appears to vary with the wave-length of the incident wireless wave; nor is it constant for any given wave-length, but varies from day to night and from season to season.

But despite these variations the atmosphere, in respect to its electric state, and its relation therefore to wireless waves, always consists of two parts: a highly ionized, conducting and wave-reflecting outer shell, the Kennelly- Heaviside side; and a relatively non-conducting, but wave-transmitting inner shell. Electrically, also, the atmosphere has structure.

Ozone Layer

The composition of the atmosphere, there is very little ozone (triatomic oxygen) in the lower air up at least to the level of the highest clouds, but certainly very much more somewhere beyond that height. From spectroscopic observations several persons have computed the height of the centre of gravity of the ozone to be 25 to 30 miles above the surface of the earth. But whatever the correct value of this Iteight, surely in respect to ozone also the atmosphere has its structure—an intermediate shell rich in ozone, and an inner one and an outer that contain practically none at all.

DISTRIBUTION OF TEMPERATURE

The pull, as we call it, of gravity makes water run down hill. It also makes a heavy liquid underrun a lighter one in the same level; both are drawn in the direction of the bottom, but the pull on the heavier, or denser, is greater than on the lighter, and the stronger pull prevails. Gravity also makes an isolated mass of liquid or gas in a heavier one go up, not down; it is pushed or buoyed up by a force equal to the difference between the weight of the lighter and that of an equal volume of the heavier. Clearly, then, whenever two masses of air of unequal density come into free contact with each other the lighter is pushed up and away, except in the case of properly adjusted winds.

Now air rapidly increases in volume, and correspondingly decreases in density, with increase of temperature roughly 1% per 5°F at ordinary temperatures. Hence the hot air in a chimney is lighter than an equal volume of the cold air on the outside, and therefore is pushed up by the latter which, in turn, is heated and itself pushed up, and so on as long as there is a fire in the grate to supply the heat.

To be sure, the combustion alters the composition of the air (makes it richer in carbon dioxide if coal is used, and in both carbon dioxide and water vapour if wood or gas is the fuel, and poorer in oxygen) in such manner as to render that in the chimney heavier, at the same temperature, than that outside, but this increase in density through change in composition is small in comparison to its decrease in density by heating. While ordinarily the effect if much less, since commonly only part of the oxygen is consumed; hence the heating, being several times this maximum value, has, in any case, the best of the argument, as it were, and the chimneys keep on drawing. Similarly, air in the open is underrun and pushed up by even slightly cooler adjacent air of the same composition, unless, as already explained, the two masses happen to be flowing past each other in the right positions and directions and with the proper velocities. Actually, the heated air expands as its temperature rises, and overflows above wherever its pressure is thus made greater than that of the adjacent atmosphere.

This overflow, or outflow, decreases the pressure at the bottom, and in the lower portions, of the heated air, and at the same time increases the pressure round about under the places of overflow—mass, hence weight, is removed from one place and added to others. This disturbs the balance. Gravity tends to restore it and thereby induces winds in the direction, initially at least, of higher to lower pressure.

If the heated region is very small, equilibrium is quickly established, unless the heating is maintained. But where the higher temperature covers a large the winds no longer flow directly towards the centre of lowest pressure but more or less round about it, owing to the rotation of the earth, in a manner seemingly most contrarious. This heating in innumerable cases is very local and of only a few hours' duration; in many others it is quite extensive and lasts days Weeks, and even all season long; while its greatest manifestation is the year after year and age after age continuously higher temperature in tropical realms and lower in the frigid zones.

This perpetual heating of the atmosphere over one great region, and its ceaseless cooling over another, or rather, two others, keeps it continuously out of balance and makes the winds, especially the trades and the westerlies, forever to blow—to blow dizzily over a rotating earth, and time and again violently and confusedly incident to the rapid, the all but explosive, delivery, by condensation, to a limited region of vast quantities of heat that had been slowly accumulated by evaporation from others afar off. The whole of the atmosphere to the tops of the highest clouds, that is, the whole of the troposphere, is a huge convection system, greatly complicated by the rotation of the earth and all but hopelessly confused by evaporation and condensation. The stratosphere, too, has its circulation, but as yet not much is known about it.

Of course it is difference in pressure at the same level that pushes the air about, or makes the winds to blow, but, as explained, this difference in pressure depends, in turn, mainly on the distribution of temperature. That is one reason, but not the only one, why this distribution is so important. Perhaps some good physicist will insist that it really isn't difference of pressure at the same height above sea level that makes the winds blow, but difference of pressure over an "isentropic surface," or surface of "equal entropy."

Well, he would be right in respect to appreciable heights above the surface, because for the free air the isentropic surface is the "level" surface. But nothing short of a surgical operation can get the idea of entropy into the other fellow's head, and there is no rivet, weld, or hermetic seal that will keep it there. Besides, commonly (not always), there isn't much difference between the two after all—"same level" and "isentropic level"—and so we will stick to the one everybody knows and no one forgets, that is, "same level."

Source of Heat

When we think of the source of heat, especially in the winter-time, we are likely to have in mind some sort of combustion, for that is the cause of the tropical climate we have indoors at that season.

But indoors is a mighty small place in comparison with all outdoors; and outdoors is heated, too, often very hot in summer, and always far above the 460° below zero Fahrenheit that would be its temperature if there were no heating at all. Almost every bit of this enormous amount of heating comes from just one source, the sun.

Incoming Radiation

The radiation from the sun is so great that if it all got through the atmosphere enough would fall on each square foot directly facing it to heat a gallon of water from the freezing point to the boiling point in just three and a half hours. But it does not all get through, and what does get through always comes in slopingly except wherever the sun happens for the moment to be directly overhead. In fact, owing to the reflecting power of clouds, especially, and the surface of the earth, and to the scattering of light by the molecules of the air and by the myriads of dust motes, one-third, roughly, of the incoming radiation is thrown off to space without producing any effect whatever on the temperature of the atmosphere or of the earth beneath.

Another one-third, again roughly, of the incoming solar radiation is absorbed by the atmosphere, and the remaining portion by the earth. These statements apply to the earth as a whole. The ratios between loss by reflection and scattering, air absorption, and earth absorption, vary widely from place to place and season to season, owing mainly to differences in humidity, cloudiness, and elevation of the sun above the horizon, and differences in the character of the surface of the earth—whether land, water, snow or ice, bare soil or vegetation.

Clear Sky Radiations

It is interesting to note that the amount of radiation reaching the earth from a clear sky is equal to a considerable fraction of that which reaches it from the sun directly. At sea level the amount of sky radiation onto a horizontal surface of any particular size, a square foot, say, is equal to about 7.8% of the amount of unaffected, or direct, solar radiation onto an equal area squarely facing the sun at the same time and place.

When the sun is directly overhead its supply of heat to a horizontal surface at sea level is nearly 13 times as great as that from the sky. When it is onethird of the way down from the zenith to the horizon its contribution of heat to the earth is only a little more than 6 times that from the sky, and each is then decidedly less than it is when the sun is in the zenith. Finally, the two sources are equal, though both are still further enfeebled, when the sun is above the horizon about one-twelfth the distance to the zenith.

The brightness of the clear sky is greatest near the sun, as even casual observations readily show, and decreases gradually with increase of distance therefrom over a large part of the whole area. Hence the total of sky radiation received per minute on a horizontal surface is greatest at noon, as is also the direct solar radiation. The intensity of sky radiation decreases, in general, with increase of height above sea level, while that of the direct solar radiation increases.

RADIATION FROM AN OCERCAST SKY

When the sky is overcast, neither its brightness nor the total amount of radiation received on a given horizontal area is at all constant, even for the same height of the nun, because the clouds in question may be of any kind from the thinnest cirrus that just dims the sun, to the darkest nimbus that reduces even noonday brilliance to twilight. If, however, the sky is completely overcast by an approximately uniform cloud layer dense enough, but not greatly more than enough, to prevent the position of the sun from showing, then the total radiation from this cloud layer onto a horizontal surface is, on the average, slightly greater than that from a clear sky. Evidently, too, the amount received of this cloud-transmitted radiation generally must increase with increase of height above sea level of the place of reception. The brightness of the cloud layer is surprisingly close to uniform. It is greatest nearly overhead (just a little way off in the direction of the sun), but still nine-tenths as bright half way to the horizon, and half as bright almost at the horizon.

DISPOSAL OF RADIATION OF SURFACE OF THE EARTH

There are just three things that can happen to radiation incident onto any extended object. It must be reflected, transmitted or absorbed. When the object is extremely small it more or less scatters incident radiation, and radiation that just grazes the boundary of an object suffers still another effect which we call diffraction. However, neither scattering nor diffraction occurs when the object is large and its edges are not involved. They do not, therefore, occur in the case of radiation incident on the surface of the earth.

Here then, the incident radiation is all used up by two processes, reflection and absorption, since there is no transmission—no passage of radiation through the earth and out at the other side. The portion reflected is about 70% for snow-covered regions, and 7% for the rest of the world.

The remainder is absorbed, that is, 30% wherever there is snow, and 93% at all other places, both land and water. That which is reflected is lost except in so far as it is absorbed by the air above. The absorbed portion goes largely to heating the upper layers of the soil or water, but not all of it, since a considerable part is consumed in maintaining evaporation, and a much smaller part in effecting plant growth and development. Another relatively small part merely melts snow and ice without raising their temperature above the freezing point. The heated surface in turn heats the soil or rock by conduction, but appreciably to a depth of only a few feet.

The heating of water extends to a greater depth owing partly to the penetration of the rays to some distance below the surface, and partly to the mixing of the water by wave action. The heated surface also warms the air above it both by direct contact and by radiation. Furthermore, the heated air through convection shares its warmth with other and colder air above; and the heat consumed in evaporation at one place is liberated, that is made sensible or temperature-producing, some other place, usually in mid-air, where condensation occurs, and far away.

Practically every bit of this heating of earth, ocean and air, and supply of energy for evaporation, plant growth, ice melting, and what not else, comes from the sun—all directly except about one part in half a million that reaches us after reflection by the full moon and the planets. A negligibly small amount comes from the fixed stars, enough to keep the average temperature of the out-doors air about two millionths of a degree Fahrenheit higher than it otherwise would be. Finally, another very small amount comes from the heated interior of the earth.

QUANTITY AND EFFECTS OF HEAT FROM THE INTERIOR OF THE EARTH

If the earth had no atmosphere, and if there were no sun or stars to send us a flood of radiation, the supply of heat from the interior (of which fourfifths, roughly, is from radioactive material) alone would keep up the surface temperatures to about 60° absolute, on the Fahrenheit scale, that is, -400°F, approximately. Hence the flow of heat from the interior of the earth per square foot of surface is sufficient to raise the temperature of a gallon of water about 1°F in 16 days, and the total flow through the whole surface out to space sufficient to heat 92,000 tons of water from the freezing to the boiling point every second of time; or enough, starting at room temperatures, to melt 1,000,000 tons of lead per second.

These are big figures, and yet all this flow of heat keeps the actual temperature of the surface of the earth only about 1/25 of a degree F. higher than it otherwise would be. The figures also tell us the surprising story that if 10,000 times as much heat came from the interior of the earth as now actually does come, or, what amounts to the same thing, if everywhere there was a sea of molten cast iron covered over with a layer of rock and dirt only 10 to 12 feet thick, the oceans above could rest thereon serene with no close approach to the boiling point, so excellent an insulator, or poor a conductor, is this material; and that if the dirt and rock crust were 20 feet thick we could go about over it ourselves in perfect comfort.

Outgoing Radiation

On the average, the earth loses to space, or emits to space, by radiation very approximately the same amount of heat each year that it absorbs of incoming radiation during the same time, plus, of course, the supply of heat that reaches the surface from the interior. We know that the loss is substantially equal to the gain because otherwise the surface would be growing warmer from year to year, and we know that this loss is by radiation as there is no other way for the loss to occur—there being no such thing as conduction to empty space. The amount of this loss, or radiation of the entire earth to space, can be estimated from our knowledge of the incoming radiation and the fraction of it that is ineffective through scattering and reflection, especially by clouds.

It can not be measured directly because we have no means of getting out beyond the atmosphere and from that ideal place pointing our heat-gathering apparatus towards the earth. But, as implied above, we can make a pretty close estimate of the average rate at which radiation is going out from the earth as a whole, and the conclusion is that it is very nearly the same as that from a perfect radiator, or "black body," at the absolute temperature 454° on the Fahrenheit scale, or -6°F. This is sufficient, per square foot of surface, to heat a gallon of water from the freezing to the boiling point in about 20 1/2 hours. The radiation from the surface of the earth often is very much greater than this value, even twice as great, or more, because the temperature of the surface frequently is far higher than the -6°F, here assumed. On the other hand, at times and places, owing to very low temperatures, it is much less. On the average, however, the radiation from the surface is much in excess of that which finally gets away to space-greater by the amount of return radiation it absorbs (nearly all of it) from clouds and the atmosphere.

The surface of the earth radiates at a relatively high temperature, hence in comparative abundance. Some of this radiation goes directly through the atmosphere, but ordinarily most of it is absorbed by the water vapour and clouds in the lower air, and a little by other things, especially ozone (when the sky is clear, for it is above all clouds) and carbon dioxide. That which is absorbed in the lowest layers is, in general, reradiated, but at a lower temperature than that of the surface. This reradiation is in every direction, half of it downward, some of which is absorbed on the way, and the rest by the surface whose initial temperature and radiation it thus helps to maintain; and half upwards to the next higher layers; and so on up and up from layer to layer, but always with decreasing absorption by the air still above and increasing absorption by that below until the entire atmosphere is left behind.

Clouds do not Check Radiation

It is a well-known fact that during still clear nights the surface of the earth, and, through it, the adjacent air, cool to a much lower temperature, especially over level land and in valleys and bowl-like depressions, than they do when either the sky is clouded, or the wind is strong. Furthermore, the lower the clouds, other things being equal, the less the cooling. That seems very simple, and would be but for one little fly in the ointment—there isn't a word of truth in it.

It might do perhaps as a dose of mental paregoric for a kid with the quizz colic, but it is no good for anything else. Just one thing alone ever reduces radiation, and that is decrease of temperature. No Clouds do not check in the least radiation from the surface below. They are, however, themselves good radiators; and as their temperature, when they are low, is nearly that of the earth, they send down to it almost as much radiation as it itself emits, and as practically all this cloud radiation is absorbed by the earth, it follows that the surface temperature remains substantially constant.

It isn't that the radiation from the earth is checked in the least, but that it receives from the cloud canopy and absorbs wellnigh as much as it itself gives out. Neither is the approximately constant temperature maintained by an appreciable wind owing to any check whatever exerted by it on surface radiation, but to the fact that the net loss of heat thus sustained, and on clear nights it is considerable, is distributed by turbulence through such a deep layer and great quantity of air that the fall in temperature is small even when the total loss of heat is large.

Temperature of Surface Air

When we talk about the "surface" air it often is advisable to explain just what air we have in mind, for this term is quite flexible. We might mean only that air which is in actual molecular contact with the surface, or that which at most is within a few inches of it, or finally, all below the height of eight or ten feet, the air to which we chiefly are exposed while outdoors. The temperature of the surface air, in any one of these senses, is determined mainly by that of the surface itself. Whatever the temperature of the surface, that also is the temperature of the contact air, and very nearly the temperature of all that air which by turbulence or otherwise is frequently brought into contact with the surface. On the stillest of nights this layer at places may be only a few thick.

During the daytime, however, especially when there is sunshine to induce thermal convection, and whenever, day or night, there is a measurable movement of the air, it is certain to be at least a good many feet thick. The essential point in this: The temperature of the air near the surface (how near varies with the circumstances) depends more on contacts with that surface than it does on the amount of solar radiation to which it may be exposed. The surface temperature of course does vary with the intensity and duration of the sunshine, and so therefore does also that of the surface air, but indirectly through contact with the surface and not directly by absorption of solar energy.

Relation Between surface Temperature and Temperature of Surface Air

As above stated, the temperature of the actual contact air must be the same as that of the surface (of the substance, ground or what not, at its surface) against which it rests. If this surface air remained fixed, as we often are told that it does, then it would seem that the air next in contact with it also should become fixed in position, and so on indefinitely. But we know that fixity of position of the air molecules does not extend to a measurable distance from any solid, for we can blow smoke past it and see the motion of the air. We therefore are forced to the conclusion that fixity of position does not apply, at least not for any appreciable length of time, even to the contact molecules. The way out of the difficulty appears to be this: The actual contact molecules of the air are at rest, like a liquid film, but they do not stay at rest. They evaporate, and as they leave the surface others condense thereon—are adsorbed—a continuous process the details of which are not yet all known. In this way the contact molecules, during the extremely brief interval of their contact, are fixed in position, but they are continuously reverting to the gaseous state, and therefore the atmosphere at ordinary temperatures is always fluid however measurably near it may be to the surface in question.

Since the air is directly heated chiefly by contact with the surface of the earth, and indirectly by the sharing of this heat, through convection, with colder air above, it follows that in general wherever the temperature of the lower atmosphere is increasing, that is, over nearly all snow-free land, and particularly during the day time, there the average temperature of the surface is higher than that of the free surface air. This is in accordance with what physicists call the second law of thermodynamics, and what everybody else knows without calling it anything, namely, that the temperature of the heater is higher than the temperature of the thing heated. Similarly, where the lower air commonly is cooled, as it is over snow-covered regions, there the average temperature of the surface is lower than that of the surface air—the cooler is colder than the thing cooled.

Maximum and Minimum Temperatures

Obviously if the heater changes temperature, so also will the heated, and the heater will be the first to change and the first to reach its extreme values— maxima and minima. There is no surprise, therefore, in the fact that the daily maximum temperature of a snow-free land surface occurs earlier, about 1 o'clock P.M., than that of the air above it, which is delayed until around 3 o'clock. The air and surface minima, occurring near daybreak, are much closer together, owing partly to the slow cooling of the soil through the night. Over the ocean the temperature of the air normally is a little higher, a degree or so, than that of the water, and the time of its maximum value, near 1 o'clock P.M., a little earlier than that of the water. This is due to the fact that here the surface air is humid and also "dusty" with salt particles and therefore absorbs a large amount of radiation, so much indeed that its daily range of temperature is more dependent on this direct absorption than it is on conduction and convection from the surface. The minimum temperatures of air and water occur simultaneously, or nearly so.

Periodic Temperature Changes

Nearly 150 different periods of temperature and other weather changes have been published, ranging from 24 hours to 744 years. Nearly all of them, however, have a shorter period than 40 years, and half of them a period of 8 years or less. Of this great number of periods there are only two, the 24-hour or daily period, and the 12-month or annual period, that everybody accepts. There is one other, the so-called 11-year or sunspot period, that is widely, though not universally accepted; and still another the 35-year, or Brückner, period that many believe to be real. No credence was ever given to any of the others save perhaps by their discoverers, and in most cases even that must have been half-hearted.

The daily period is everywhere conspicuous (save for part of the time in polar regions, when the sun is continuously above or continuously below the horizon), and in respect to temperature, gives, on the average, a maximum in the early to mid afternoon and a minimum shortly before sunrise. Over the oceans this diurnal range is only 1°F to 3°F as a rule. It also is small in the humid and cloudy portions of the continental tropics, owing to the large amount of return radiation from the clouds and water vapour.

In desert regions, especially at high altitudes, where the sky is clear and the humidity very low the diurnal range of temperature is at its maximum. In extreme cases this range is of the order of 100°F, from distinctly below freezing to decidedly over 100°F—both in the shade. The annual range also is extremely conspicuous in most parts of the world. In this case the exception does not occur at and near the poles, but at and for some distance on either side of the equator. At the equator the "year," as it were, counted from the time the sun is overhead at noon until farthest away (23 1/2°), and then back again is 6 months.

Next beyond the equator on either side there obviously are two such "years" but of unequal length. At one distance they are 5 and 7 months, at another 4 and 8, and so on until at the Tropics, Capricorn and Cancer, only one is left, and that one 12 months in duration, the same as from there on to the pole. The times of occurrence of the annual maxima and minima vary widely from place to place, but always they are after maximum and minimum reception of heat from the sun. The delays are least over inland deserts and greatest over mid to high latitude portions of the oceans.

The sunspot period, approximately 11.1 years, is most pronounced at high levels within the tropics. Here the average temperature during the year or two around sunspot minima, or when the spots are fewest and smallest, is about 2°F higher than the average temperature during the time of spot maxima. The same relation appears to hold, in general, for the middle and higher latitudes but with decidedly less contrast.

Just to make the list complete it may be worthwhile to mention a few utterly unimportant but entirely real temperature periods. Those are the periods of the changes in light and heat received from the moon-maximum at full moon, minimum at new moon; changes in the distance of the earth from the sun due to the pull of the moon in its orbit about the earth; and similar but far less changes of and by the planets. The sum total of the effects of all the planets is about equal to that of the moon alone, that is, a change in the average temperature of the earth of about .02°F, due almost wholly to variations in our distance from the sun or as we say, to perturbations in the earth's orbit. But, as already stated, this change in temperature is too small to bother about.

Temperature Lag

It was stated above that the hottest time of the day is not noon, when the sun is most effective, but two to four hours later; and similarly, that the coldest weather does not come with the shortest days, but generally a month or so later. In proverb form: "As the days grow longer the cold grows stronger." In the early morning of a clear day following a cloudless night, say, the earth and surface air are relatively cool. Then with sunrise they begin to warm up, but not rapidly, even when there is no wind, because it requires an appreciable amount of heat to warm even a pound of soil 1°F, and several times as much to equally warm a pound of water.

But as the sunshine continues, the soil at first gets hotter and hotter, and as its temperature rises the rate at which it loses heat by radiation rapidly increases. However, since the soil, including of course its covering, warms slowly, owing to its large capacity for heat, its loss by radiation falls more and more behind its gain by absorption as the sun rises higher in the heavens, and therefore catches up with the latter only in the afternoon when the insolation is distinctly less than it was at midday.

Hence the diurnal maximum temperature, whether of the lower air or of the surface of the earth (an earlier phenomenon) necessarily lags behind the maximum intensity of the sunshine. Similarly, the annual maximum temperature occurs several weeks after the days are longest and the heating strongest. Very similarly too, because the earth can give off stored up heat when the supply becomes deficient, the minimum temperature comes several weeks after the shortest days. During this period, as the days grow longer the cold grows stronger. The diurnal and annual heating and cooling, and lagging of temperature extremes, may be likened to the alternate rise and fall of the water level in a reservoir having a continuously open drain pipe at the bottom and a periodically variable inflow, now greater, now less, than the then rate of outflow, but so regulated that the reservoir may never become empty.

Day Degrees

Not only are we interested in the values and times of occurrence of maximum and minimum temperatures but also, and even more, concerned in the occurrence of certain critical temperatures. For instance, we are very much interested in the temperature at which frost can occur until it does occur, after which, if it has been a "killing" one, we are no longer much concerned as there is nothing left for the next one to injure. Another critical temperature is 42°F as that closely marks the boundary between growth and dormancy for most vegetation of the temperate zones. In fact it is customary to call the difference between the average temperature of a given day, if higher than this value, and 42°F, its day degrees. The sum of these daily values over a week, month or season, is the number of day degrees for that period, and is an important index to what might have been the vegetable growth during the time in question. Similarly, engineers and others interested in artificial heating of buildings, count day degrees relative to a temperature of 65°F.

Wind Direction and Temperature

The effect of wind direction on the temperature of a place depends on its location. Well within the Tropics, and also near the poles, the effect of wind diıection obviously is small because the temperature is pretty nearly the same round about in every direction. In middle latitudes, however, the situation is quite different, partly because here the temperatures commonly are not the same in every direction, and partly also, in fact mainly, because here each section of the cyclone and of the anticyclone has its own wind direction, and some of them a wind system entirely distinct from that of the others. In the forward or eastern portion of the anticyclone the winds are from the region of higher latitudes, and having come a long ways often are distinctly cool to cold for the place and time of year. Similarly, the winds of the western segment, having come from much nearer the equator, usually are relatively warm. In the cyclone, or widespread disturbance, all that segment of 90°, more or less, lying between one line running east, to southeast, from the storm centre and another generally south to southwest (in the northern hemisphere; east to northeast, and north to northwest, in the southern hemisphere) is occupied by a great current of warm air from low latitudes. The rest of the storm area is covered with cold winds from the east, north, and northwest, in succession as one in the northern hemisphere passes from the front to the rear of the storm centre on the poleward side; from the east, south, and southwest, in the southern hemisphere.

In general, all these cold winds in each hemisphere are of polar, that is, high latitude origin. Clearly then, the temperature of the air in a cyclonic region is likely to change with the direction of the wind. In one portion of this disturbance, namely, along a narrow strip that meteorologists call the cold front, or wind shift line, and which commonly runs west of south (west of north in the southern hemisphere) from the storm centre, the wind direction rapidly changes from southwesterly to northwesterly, with, as a rule, a sharp drop in temperature as the tropical breezes give way to polar blasts. Hence in middle latitudes air temperature is closely dependent upon wind direction, both in cyclones and anticyclones; and that means the greater portion of the time, for usually we are in the midst of one or the other of these disturbances.

LOCATION AND TEMPERATURE

Every one knows that the average temperature of the tropical regions is higher than that of the polar areas, but it is not a familiar fact that, nevertheless, int he course of a year the temperature reaches 90°F or over (in the shade) on more days in central Alaska than at Panama. It also is a surprising fact to most of us that the average temperature through January at St. Louis is the same as that in southern Iceland; and that the average temperature for the entire winter, December, January and February, at Sitka, Alaska, is about the same as that at Washington, D. C. And there are lots of other similar surprises, as, for instance, the fact that semi-tropical vegetation that would be killed by frost is now growing wild on the Scilly Islands in the latitude of northern Newfoundland. Clearly, then, difference in longitude may be accompanied by nearly as great a variation in temperature as is difference in latitude. During winter especially the lines of equal temperature run far poleward over the oceans, and equatorward over the continents.

Another matter of great influence on the local temperature is the nature of the surrounding area, both near and distant. A fair inland point, for instance, becomes much hotter in summer and greatly colder in winter than does a mid-sea island at the same latitude. Also a coast where the prevailing winds are on-shore has a more nearly even temperature than one of the same latitude with off-shore winds. The first is bathed in ocean breezes of relatively equable temperature; the second in winds that have traveled far over land and that therefore are characterized by its temperature irregularities and extremes.

It is interesting in this connection also to note that the temperature is a little higher, and in some cases quite noticeably higher, in a city than in the adjacent country; and further that in the country the forest is cooler in summer, and slightly warmer in winter, than the open fields.

In some cases the foot of a high mountain occasionally is very peculiar in respect to temperature. It may happen that the air is quite cold when all of a sudden there comes across and down the mountain a roaring hot wind that within a few hours clears away every trace of even a deep snow. This wind went up the other side of the mountain saturated and rainy, hence it cooled relatively little with ascent owing to the latent heat of vaporization there rendered sensible by condensation. As it came down, however, it was dry and therefore heated rapidly with descent and consequent increase of pressure. Such are the famous foehn winds on the northern side of the Alps, and the chinook winds of the Rocky Mountains.

South Pole Colder than the North

We often are asked which is the colder, the north pole or the south. One answer might be that, as we have no records longer than a few hours at either, we don't know. That is true enough so far as bare statistics are concerned, but in this case we can reason the matter out from sure and simple premises:

- Normally, the greater the height the lower the temperature, other things being equal;
- The south pole is 10,000 feet, roughly, above, and the north pole at, sea level. Hence we should expect the south pole to be the colder of the two.
- The greater the height the less, in general, the amount of cloud, water vapour and other gas to radiate back to the earth and thereby help to keep up its temperature;
- The south pole is much higher than the north. The south pole, therefore, because it gets less return radiation than does the north, should have the lower temperature.
- The faster heat is supplied by the surface to the lower air the higher the temperature of that air, other things being equal;
- By actual measurement the ice over the Arctic Ocean gives off enough heat in 24 winter hours, coming from the relatively warm water below, to increase the temperature of a layer of air 450 feet thick by 20°F, while that given off at the south pole is only a small fraction of this amount, owing to the much poorer conductivity of the snow and the far greater depth to a temperature equal to that of the arctic water.

Then for this reason also the south pole must be colder than the north. It seems therefore that even without the actual observations we may feel reasonably certain that the temperatures of the two poles are not the same, especially their night temperatures since these occur when they can be but little affected by the variations in our distance from the sun, and that the south pole is distinctly the colder of the two.

Relation of Surface Temperature to Height

Those who live in mountains regions know by personal experience that, in general, the higher up the mountain the lower the temperature. Instrumental records show that this relation is true not only for mountains but also for hills and even plateaus, and that the approximate numerical values are 1°F decrease per 330 feet ascent on a mountain, 365 among hills and 455 on plains.

The reason for this difference in favour of the plateau is the fact that the air is heated mainly by the surface of the earth which, in turn, is heated by the sunshine. That is, in the case of the plateau the surface which is the heater of the air is at the height in question all around as far as the level area extends, while the air on the mountain is affected in part by the temperature of the free air, especially when there is an appreciable wind, and this free-air temperature is lower, except on still clear nights, than surface air at the same altitude.

Relation of Temperature to Height in the Free Air

According to observations the temperature of the air normally decreases with increase of latitude from the surface of the earth up to the height of several mile, roughly 6 to 7 in middle latitudes. Beyond this level the temperature remains substantially constant up to the greatest heights yet attained, probably around 18 miles.

The average rate of decrease of temperature of the free air with increase of height is about 1°F per 300 feet, from the surface up to the level at which appreciable decrease ceases—up to the "tropopause," or limiting reach of convection.

In general this temperature decrease is most rapid in the upper half to two-thirds of the depth under consideration and least rapid in the lower onethird. Immediately above a land surface the change of temperature with height is widely variable, from a decrease, perhaps ten fold the above average value, to an increase ten fold that rate, at least through the first 20 feet or so, above suitable regions, such as plains, valley bottoms and bowl-shaped depressions, and, of course, during still clear nights.

Few phenomena of the atmosphere are as often "explained" as is the fact that, in general, temperature decreases with increase of height, and hardly any other as inadequately, not to say erroneously, explained. The facts are:

- Half, roughly, of the sunshine that is not lost by reflection or by scattering, that is, half of it that does any heating at all, gets entirely through the air and is absorbed by the surface of the earth, which, on being thus heated, heats in turn the adjacent or lowest air.
- The other half of the effective radiation from the sun (portion used and not immediately lost) is absorbed mainly by the water vapour, and as the density of this vapour generally decreases very rapidly with increase of height it follows that by the direct absorption of sunshine the heating of the air likewise is greatest in its lowest levels.
- The surface of the earth loses heat (it doesn't keep on getting hotter and hotter indefinitely) not only by conduction to the adjacent air, but also by radiation, a kind of radiation greatly absorbed by water vapour. Therefore in this third way, too, as in each of the others, the atmosphere is more and more strongly heated with decrease of height.
- Although the surface air is most heated it is not equally heated everywhere. Hence the warmest and lightest portions are pushed up—forced to rise—by the cooler and denser air round about as a cork is bobbed up when let go of under water.
- The ascending (pushed-up) warm air comes under less and less pressure with increase of height by the weight of the air left below. It therefore continuously expands while rising, and all the time against pressure—the weight of whatever air is still above it. But this expansion against pressure is work, and work at the expense of the only supply of energy the rising air has—its heat. Hence as it rises it must and does become cooler.
- This cooling of the air by convection does not go on to absolute zero, nor, as explained, does the surface air keep on getting hotter and hotter. And these limitations are owing to the fact that the atmosphere loses heat by radiation. The free air therefore is all the time gaining heat by absorption of radiation and losing heat by emission of radiation, while the surface air is gaining heat also by contact with the warmed earth.

- But loss of heat by radiation decreases very rapidly with fall of temperature, while the power to absorb radiation does not change. Hence as the air ascends higher and higher, and thereby gets colder and colder, it presently comes to a temperature at which its loss by radiation is equal to its gain by absorption. Beyond this level it can not rise, because if it did so it instantly would become colder and denser than its environment and fall back again. From this level on up the temperature of the air must remain roughly constant except as modified, perhaps, by change of composition. This is the isothermal region, or stratosphere, which every planet must have whatever the extent and composition of its atmosphere. 52 Encyclopaedia of Environmental Science
- Since the cold air of the stratosphere is losing heat by radiation at the same rate that it is gaining it by absorption it follows that the warmer air of lower levels is losing by radiation faster than it is gaining by absorption, the net difference at each level (not moment to moment for the moving air) being made up by heat brought there by convection from the surface, either directly as such, or indirectly through evaporation and subsequent condensation. In this way, that is, by heating below and cooling above, convection and, in general, a decrease of temperature from bottom to top of the convective layer, is maintained without the air as a whole getting either warmer or colder.

Temperature Changes in the Stratosphere

Although the stratosphere, that portion of the atmosphere beyond the highest clouds, has no immediate contact with the warming and cooling surface, nevertheless its temperature at any given locality often varies 10°F to 20°F, and even more, not vertically, as a rule, but horizontally, or from day to day. Thus the temperature of the stratosphere commonly is distinctly higher over the forward and central portions of a cyclonic area than it is over the corresponding portions of an anticyclone. The cause of this change is not definitely known.

Temperature Inversions

The structure of the atmosphere, during still clear nights the surface air becomes so cold over level land areas, in valleys and in basins, that often there is a rapid rise of temperature with increase of height through the first 10 to 100 feet, or more. This is the surface inversion, so favourable to production of frost, and without which orchard heating commonly would be unnecessary and, moreover, in general impracticable. Every wind of appreciable strength so thoroughly mixes up the lower air by turbulence that the temperature of the top portion of the agitated stratum is decidedly lower than that of the undisturbed air immediately above it. This is the turbulence inversion, which, because of its low temperature often is accompanied by a stratiform cloud.

It might seem that a similar inversion should occur at every interface between over- and under-running air currents, but such inversions, so far as they exist at all, are too slight to be of any particular importance. This is owing to the smallness of the friction between free air currents and the consequent all but complete absence of turbulence.

The highest temperature inversion in the atmosphere, of which we have any actual and direct measurement, is that at the base of the stratosphere during the passage of an anticyclone. As already stated the cause of this particular, and often very pronounced, inversion is not yet definitely known.

THERMAL BELT, OR GREEN BELT

There are various ways of protecting fruit from frost, but the best of them all is the proper selection of the orchard site. In a hilly or mountainous region that best location is neither on the floor of the valley nor, generally, on the top of a ridge, but, as a rule, some distance, not too far, up one or the other of the slopes. A strip along a hillside or mountainside at this level is known as the thermal belt, green belt, verdant belt, frostless belt, etc., because on still, cloudless nights this level is warmer than any other above or below it, hence least likely to have frost, and most likely to show green and uninjured vegetation.

After sundown, when the sky is clear and there is no wind, the surface of the earth everywhere cools much more rapidly than the free atmosphere, and in turn correspondingly chills the nearby air either directly through actual contact, or indi- rectly by turbulence mixing with that which had been so chilled.

On the side of a hill, then, this air, because it is denser (being colder) than that at the same level over the adjacent valley, flows down slope much as would a sheet of water. However, as this drainage air reaches lower levels it evidently is subjected to increase of pressure to the extent of the weight of the air passed below.

It therefore is compressed—work is done upon it—and its temperature made higher than it otherwise would have been. In the early evening, and well up on the hill, this heating of the down flowing air causes it to become warmer and warmer with descent, almost to the valley bottom. Here the gain of heat through increase of pressure not only makes up for all that was lost by contact with the cold surface, but actually warms the air to a higher temperature than it had before it was first chilled. But this heating is pronounced only where the descent is rapid. Air already at the bottom of the hill is not thus heated. It does not come under any greater pressure for there is no lower place to which it can rapidly drain. It therefore just gets colder and colder as the surface temperature continues to fall incident to the net loss of heat by radiation.

It also gets colder, but not so rapidly, near the bottom where the slope is gentle and the current sluggish; and less and less rapidly with increase of height and gain in the speed of flow. There must therefore be some level along either valley wall at which, for the time being, the heating of the descending air by compression is just equal to its cooling by contact with the chilled surface. Above this level the air evidently gets colder with ascent and below it colder with descent. During most of the night the flood of cold air grows gradually deeper, carrying the level of maximum temperature—the level of its crest-higher and higher up the valley sides. Near morning, however, it becomes practically stationary, and where it then is frost obviously is least likely to occur. This is the place to plant your orchard—here along the thermal belt where the temperature is highest and the chance of frost the very least; where vegetation may pass through the night unharmed, while all above is frozen stiff and all below white with frost.

❐

Index

❑❑❑